# A
# CANADIAN
# JEW
# IN
# UNCLE SAM'S
# ARMY

# A CANADIAN JEW IN UNCLE SAM'S ARMY

---

Darkly Humorous Recollections
of Disparity, Irony,
and Narrowly Surviving
the Cold War

---

## ADAM L. HARRIS

EMERGENT ISLES PRESS

Published 2024
Printed in the United States of America
Hardcover ISBN: 979-8-9896832-0-8
Paperback ISBN: 979-8-9896832-1-5
E-ISBN: 979-8-9896832-2-2

Emergent Isles Press
Dowling, Ontario, Canada
emergentislespress@gmail.com

*Book design by Stacey Aaronson*

Names and identifying characteristics have been changed to protect the privacy of certain individuals.

Droll thing life is, that mysterious arrangement of merciless logic for a futile purpose. The most you can hope from it is some knowledge of yourself that comes too late, a crop of unextinguishable regrets.

—JOSEPH CONRAD, *Heart of Darkness*

# INTRODUCTION

⚊⚊

I started writing these stories down when they began kicking holes in my head to get out. Actually, the holes were already there and it's the details that began to fall out. Just into my sixties, I realized I could no longer remember the names, the people, as I once could. I still see their faces, their words still echo in my ears, but they are slipping away as surely as I am.

Like so many soldiers, my writing began as the result of experiencing the isolation of Basic Training, and then the unpredictability of Army life. Letters home became a way of maintaining a bridge to a far saner, safer world of the families we were guarding so many miles away.

Those letters I wrote were given back to me by my brother, Harvey, ten years after they were mailed. It was the last time I ever saw him. They sat undisturbed in their sealed Manila envelope through seven moves, for almost three decades. Though heavily edited—I had to cut out most of the profanity, which often didn't leave much—they are real. But what I didn't write home about, what was once only written between the lines, is in these chapters.

Most accountings of the Cold War are either told by ex-spies, retired generals, or historians. Rarely do you get a lowly private's-eye view. The lives we led to break the back of the Soviet Empire so severely that it got reduced to being Russia again has never been truly told until now. What we, the lower enlisted, were put through would have been illegal to do to criminals. But when injuries go unreported, deaths never make the news, and an entire division gets dissolved, it's easy to bury the stories with their dead.

It's true that my time served may have been fairly contentious, and I do have the dubious honor of being discharged as a lower-ranking private than when I started, but the closeness and camaraderie of the service is a heart-wrenching gift of redemption I will carry untarnished forever. This I write for the brotherhood I served with, bunked with, ate with, and drank with, the guys who, like me, took the orders and did the dirty work—the American GIs.

# ONE

~~~

## Why Resurrection Isn't for Everybody

*"...the most terrifying question of all may be just how much horror the human mind can stand and still maintain a wakeful, staring, unrelenting sanity."*

—Stephen King, excerpted from *Pet Sematary*

"**H**arris! See that 5-ton? Grab your gear and go—pronto!"

An 8-inch howitzer had a broken road wheel and the crew needed help replacing it. It was the Cold War, and American field artillery troops spent a lot of time in Grafenwoehr, Rommel's Playground, perfecting their craft. It was here, deep into Graf, where I jumped into the back of a five-ton truck with a handful of other cannon crewmen, having no idea my life was about to irrevocably change.

To the civilian eye, the howitzer looked like a tank on steroids; to compare the 8-inch—so named because its projectiles were eight inches across—to a battle tank would be like comparing a stretch limo to a Mini Cooper. The Army had constructed this brute of military engineering by taking a mammoth gun barrel off a destroyer battleship and mounting it on an oversized set of tank tracks, creating the military's largest self-propelled howitzer. With pinpoint accuracy and a top speed of thirty-five miles an hour, the 8-inch can drop a two-hundred-pound round into a fifty-gallon bucket from fourteen miles out, then scoot to safety in less than a minute.

Like a tank, the 8-inch has a set of road wheels, like monstrous cogs, that engage the track from within and force its 62,500 pounds of metal into movement. Five hulking wheels, and one compact one, fit into the tracks on either side of the vehicle, all powered by a Detroit super-charged diesel V-8 engine. To change out a road wheel means opening up the track by removing a lynchpin big enough to be an imp's walking stick.

Practicing for war could be a deadly exercise in itself. In this case, we were dealing with wheels weighing more than a thousand pounds each, so none of this task was undertaken lightly.

When we got to the broken track vehicle, we were met by the APC—armored personnel carrier—that was delivering the brand-new wheel. This was just one of many stops for the five-ton we were in, so I grabbed my rucksack I'd been reclining on, and the five of us in the back of the truck reluctantly climbed out and slogged through the thick muck to the carrier. Our job was to carry the spare to the broken-down howitzer. Though somewhat grueling, it was yet another menial task on yet another gray day in the Schwarzwald.

It was mid-afternoon, but time is lost in a place where the sun is banished and it's bone cold from morning to night. A fine ceramic-like clay—the result of over a century of heavy field artillery tracks pulverizing the bedrock—coated everything in the same pallid color as the overarching mood. When the clay was wet, it became an especially thick and viscous mud, so Grafenwoehr was either a glutinous mess, or a stifling and gritty earthy-smelling wasteland with nothing in the middle.

While we waited for the spare, I had to relieve myself, so I walked off and hung my sack on a squat tree to do my business. The downed gun crew was having difficulty removing their track for the switch-out and called three of the other guys over to help. The APC was parked just off the road with its deck angled down toward us, about shoulder high, and I rejoined the lone private waiting at the rear of it.

The two guys holding on to the wheel had propped it up for us. But they'd gotten tired, so they decided to lean it on the vehicle floor. As they did, the fine drizzle that had slickened the smooth aluminum as well as the hard rubber casing on the wheel made them lose their grip. Like a greased projectile, the huge metal disc slid across the carrier's floor and fell like a coin from a slot, hitting my right foot with such force that it leapt a yard off to the side. The muddy ground beneath my foot helped absorb the impact, but the whole thing happened so fast that the extent of the injury, encased in my boot, wasn't immediately apparent.

"I need to go see the medic," I vaguely recall saying, retrieving my bag and limping away.

Shock can be so gentle, so kind. The last thing I remember is being told by a blurry medic to go back to my platoon because we were bugging out . . . and then nothing.

— — —

Seven days later, about an hour before sunset, a young second-lieutenant stumbled upon my tent.

"Holy fuck," he uttered. He flinched at the multi-hued glob at the end of my right leg where a foot should have been. "Who are you?"

When our eyes locked, I jolted awake as if I'd been sleep-walking for a week. I found myself propped up by two pine crutches, holding a pine-branch broom in one hand. (Tidiness was essential for keeping wild boars from getting too close.) I had no recollection of what I was doing there. I only knew the guy staring me down looked completely terrified.

The serviceman took a few tentative steps closer to me. "How have you been eating?" His eyes darted about wildly. "How are you alive?"

When I had no response to offer, he got panicky. "Stay put," he ordered, then ran back into the forest the way he had come.

In less than half an hour, a helicopter landed in the small clearing outside my tent. The team immediately saw that I was far from lucid and medevacked me so fast that all my gear—tent, cot, and sleeping bag—was left behind. Even in my near stupor, I couldn't help but think about how the Army felt about excuses, and how one way or another, I was going to have to replace all three on my dime.

———

At Rhein-Main I was examined by befuddled Army docs, and then by equally befuddled civilian MDs. None of them had ever seen anything like my injury. Every cell in my foot had ruptured, resulting in the comic-book edema that had spooked the poor kid who found me. Yet the skin hadn't split and not a single bone had broken. If either of those things had occurred, infection would have set in and I would have never lasted the week. But by a sheer stroke of luck, the weight of my boots, along with the twelve inches of muck I'd been standing in, had saved it.

Though my foot no longer resembled a human appendage, the civilian doctors decided that if I'd made it that far without any medical intervention, they would adopt a "let's wait and see" attitude. The Army doctors followed along, which suited the current military dogma. With it being the height of the Cold War, and political tensions high between the US and Russia, they figured that with or without the foot, I could still fire a fifty-cal—unless, of course, I was on pain medication. So, stuck between these two sets of MDs, I received no pain meds and no other care beyond what the nurses did to make me comfortable, which was to periodically adjust my tensor bandages to help that thing that used to be my foot gradually regain its shape, and thankfully, to help hide it.

After five days, I was put on a small bus full of other injured GIs and dropped off at the front of my post in Hanau, where my "job" became staying in bed and resting up from nine to five. After that, I was on my own.

My size eleven was so grossly misshapen that my toes were no longer visible; they'd been swallowed up by a turgid sack of fluid-filled skin. At any given moment I could count at least nine different swirling colors, with shades of stormy purple and angry red being most dominant. With every heartbeat, lightning pulsated through my foot. The agony was so intense that I began to fantasize about amputating my foot just to disengage from it.

Instead, I started drinking.

If you're hurt bad enough, no amount of alcohol will kill the pain. But if you drink enough, after a while you just don't give a damn.

My physical therapy amounted to hobbling across the street to the bar and drinking enough to tolerate the misery of making my way to the dance hall. Even with proper crutches it was an arduous chore. Every night, I would drink until I could dance, if you could call it that, then stay till they closed the place down. Then I'd hobble back to the post to be in my cot by five in the morning and sleep the day away before doing it all over again.

That week of being lost and forgotten in the Black Forest with a pulverized foot never actually materialized for me. I only recall being found. It seemed clear I'd been deserted by my unit and had pitched my tent and assembled my cot. I had then subsisted on MREs (meals ready to eat), and even regularly swept out my humble abode. No memory of those days exist for me, though. They are deep-sixed within my subconscious, which is probably a blessing.

At one of my checkups, I asked an Army doc how I'd lost an entire week without a head injury, and why it always felt like something was crawling up my right ankle.

"I'm no shrink," he said. "You should talk to the psychiatrist."

I knew better, though. There was no way I was going to get remanded to the psych ward and be bounced out on a Section 8, so I stopped asking. Maybe the mental vacuum of that week was protecting me from something, and frankly, I'm okay with that.

After my battery returned from the field, my top sergeant came to see me, a set of papers in his hand. I knew all too well what they were: a reprimanding Article 15.

"What happened?" I asked Top. "Why didn't anyone come find me?"

He looked down at my foot. His eyes lingered as he slowly stuffed the papers into his front pocket. It seemed he had come by to make sure I wasn't shamming.

He cleared his throat, then told me that my platoon sergeant, DumbOx, had listed me as AWOL in the field. That's why they didn't bother looking for me.

*So much for no man left behind.*

The irony of it all was that the only person even considered for punishment was me.

It's usually tough to pinpoint the exact moment when one's

allegiance gets turned, the sobering realization that one's fidelity is rewarded with zero loyalty in return. But the moment I heard my top sergeant's excuse for leaving me mangled in the forest to die was when the seed of my discontent firmly took root. I knew DumbOx didn't like me, but I was gobsmacked to think he was so vindictive he would leave me for dead. And my top sergeant went along with it? My superiors clearly didn't give a damn about me, and if they buried me along with the road wheel incident to cover their own ineptitude, then so much the better. The thought that pounded my mind was: *my survival must have been a staggering inconvenience.*

With my orders for bed rest, I had plenty of time to lie in my cot, with my foot suspended like a glaring onlooker, and stew over the fact that they would have had their perfect alibi had my top sergeant followed through with issuing me the Article 15. I felt betrayed and let down in a way I didn't think possible. And it made me angry, really angry.

The Adam who came back from the dead had a whole new don't-fuck-with-me attitude. I was not only going to rock the boat. If I could, I was going to tip it over.

# MISERY

Misery, oh misery,
All my life is misery.
I used to sham from 9 to 5,
Now I fight to stay alive.

Misery, oh misery,
All my life is misery.
I used to drive a Cadillac,
Now I hump my rucksack.

Misery, oh misery,
All my life is misery.
I used to date a beauty queen,
Now I sleep with my M-16.

Misery, oh misery,
All my life is misery.

# TWO

## Song of the North

**B**read 'n' buttered in the Ottawa Valley, Canada, I was named for my Great Uncle Ansel. He was also known as the Crazy Butcher, which is why my name is Adam and not Ansel. The family always gathered on Friday nights at my grandparent's farm for the sabbath dinner. The night I was born was no exception, two weeks shy of the 1963 New Year.

From the moment I entered this world, I was challenged by normalcy. As a newborn I was completely silent, a rarity worse on the nerves than it sounds. I didn't laugh, cry, babble, or coo, to the point that my mother was driven to distraction. There were baby monitors throughout the entire house in case I did let out a peep, but I'm told I never did. Not a lot is known about "silent baby syndrome." You can't ask a baby why they don't make a sound, so it remains a mystery. It didn't help that I was also "tongue-tied" to the nth degree (ankyloglossia is the medical term for the tongue being attached to the bottom of the mouth). As with all things there are the occasional extremes. Such was my case.

Raised by my single mother with two sisters, a year up and a year down apiece, I never did know my gambling father. He had been engaged to the daughter of a French-Canadian mobster, gotten cold feet at the last minute, and skipped town the night before their wedding. Looking for a place to hide out, he found our small Jewish community, courted my mother, then converted and married her. Bankrolled by wedding gifts, he promptly disappeared on a binge, leaving my mother already pregnant with my elder sister.

As far as I can tell, my little sister and I are products of visits from him to secure funds for future "ventures" and hopes of my mother's that he would finally settle down and stay home. The guy would beg, borrow, or steal just to get back to the tables. He even stole my uncle's credit cards. A classic "flowers one day and gone the next" rogue, I don't remember him at all; there are no pictures of him and my mother never spoke about him. Just a black hole of longing in my young life, and maybe hers too.

Most days I spent my time on my grandfather's farm, for whom my mother kept the books. No one knew the surrounding fields like me: where the rats nested in the old, rusted cars; the best place by the creek to catch tadpoles; the blackberry bramble only the ants, birds, and I knew about. I learned how to watch plants and animals long enough to understand what they were saying. I might not have been able to speak the language of man, but I could imitate just about every creature I came across.

The farm and the fields were my safe haven, where I could simply and quietly be part of the nature that surrounded me. I would spend from morning till sundown outside, where there were crabapples aplenty, and wild rhubarb and sweet-tart raspberries just waiting for me to snack on, with a jar for tadpoles in one hand and my pockets bulging with yellow fodder peas to load my shooter with. You could walk out into that bordering forest, keep going and never turn back if you didn't want to, and most days I didn't want to.

My earliest memory is at the farm, lying across my Grandmother's lap, four years old and shirtless, having come from playing in the fields. I can still smell her perfume, still feel her fingertips slowly tracing patterns across my back, simultaneously creating and satiating an exquisite itch as I drift away. She died shortly thereafter, when I was too young to understand.

My schooling had started at the Hillel Hebrew Academy—which my grandfather was a founding father of—with my sisters and cousins, but I never made it past the second grade. Our teacher Mrs. Gringorten would put me up in front of the class daily, forcing me to recite the day's lesson, garbled out loud as best I could. It was her strong belief that I simply had a psychological blockage and that my inability to speak could be cured through constant humiliation. My Uncle Stan just happened to be the principal there, and while walking past the classroom one day, he witnessed the ordeal. Why my uncle and my mother didn't sit that bad bitch down, read her the riot act, and tell her to leave me alone I'll never know, but they didn't.

Every kid I knew went to Hillel; it was safe and I could weather the Gringorten storm. But because of my speech impediment, it was decided that if I couldn't even speak English, then why have me in a Hebrew school? I was out of Hillel that afternoon and enrolled in Hilson Avenue Public the next morning.

I vividly remember that first day: I was so scared that it took both my mother and the teacher to wrestle and pry my hands and feet from the door frame to get me into the new classroom. I may have been young but I wasn't stupid, and I knew that I was going from the frying pan into the fire. Not only was I the new kid at Hilson, I was the only Jew. Since there were no Black, Asian, or Hispanic kids, that made me the closest thing to a person of color. In an ocean of white Anglo-Saxon Protestants, I stood out as the

"Isle of Other," never to be part of the reindeer games and at best the lone Indian against all the cowboys.

I shared, with all others found in such dire circumstances, the lessons of being the one and only. In my case, I was "The Jew-Boy." I'll never forget the first time I was called that to my face in class; it was in the third grade and the brand stung so much it brought tears to my eyes. In that instant I realized that was how they saw me, and that's all I was or would ever be to them.

Every morning of every grade at Hilson, they said the Lord's Prayer while I alone stood silent, a daily reminder that I wasn't one of them, that their God wasn't mine, that I was different. I remember helping a girl I liked cross the street during Christmas-time, and her telling me that I had killed her God. What does a kid say to that? And who teaches a kid to say that to others? What a mind fuck that was. In her insular world, there was no one else she could say that to, so I figured she must have been saving that dagger just for me.

I was also the youngest and shortest in my class, and maybe even a little chubby (my pants' size was named after the loyal and hard-working "Husky"). But the icing on the cake was my severe speech impediment. I was completely unintelligible to all except my mother, and I think she was faking it most of the time. It's not that I couldn't talk; it's just that I sounded like a drowning animal.

It wasn't until I was almost eight that they could perform the surgery to free up my tongue, and by then the doctors weren't sure to what extent I would be able to talk. The speech center of the human brain is a tender thing; if not nourished early, it often does not develop at all, as evidenced by cases of "enfant sauvage"—and there was no doubt that I had a wild side.

But I am very much the product of a woman who refused to

be "the mother of the freak," hence being sent to aggressive therapy sessions that were beyond frustrating and maddening. More
often than not, I hated it, not understanding why I had to go at
all. I was fine with being quiet. Turns out that therapy was about
more than just the ability to communicate. It was about engaging
others, which was something I didn't have a lot of interest in. But,
thankfully, my Mum's perseverance won out. I did have a wonder-
ful speech therapist (who taught me how to lick peanut butter off
the tip of my nose and to whom I owe many thanks for the ability
to enunciate). Ironically, it would take the US Army years later to
teach me how to use my voice without fear.

When you grow up with a serious speech impediment you
don't talk on the phone, you don't put your hand up in class, and
you avoid attention. Questions you have you write down and re-
hearse saying, but never ask. The teachers know not to call on
you, so you just sit in the back and stare out the window.

In the mornings my mother would drive my sisters to the
Hebrew Academy, and I would walk to Hilson. Getting there was
the easy part; getting home was trickier. From the get-go I knew
that I had no backup. It was just me.

To say I fought every day is a lie of omission—I got beaten
up every day. In the beginning I ran, but home was too far away. I
was never fast enough, and the bullies would overtake me from
behind. Turning my back created the potential for the outcome to
be that much worse, so I quickly realized that if I was going to get
knocked down, I might as well see it coming. It wasn't until another
such beating deep into the winter of third grade that desperation
proved to be a real mother for inspiration.

I needed a way to fight back against overwhelming odds.

I needed to level the playing field.

# THREE

Nature and Nurture

Adam at times plays with many friends but keeps coming back to one friend. He feels that he must protect this friend even at the risk of getting in trouble. Adam is very sensitive to the name calling. It is hard to tell a little boy to forget about name calling, thus he gets into scrapes on this account.

—Ms. K. Lancaster, 3rd grade Report Card (age 8)

I t was after getting beaten up and then being humiliated by having my face rubbed raw in the snow after school one winter afternoon, walking home with the world darkening all around, that I formed my battle plan. Beating me up had become a spectator sport, a savage form of amusement for the others. I had no chance against the principles, the ones really kicking my ass, but then and there I realized that I could target those on the periphery— the eggers-on, the hecklers. Anyone watching, anyone at all, would be fair game.

I knew I couldn't confront them after school because they'd have backup. They always did. But I could go after them when they weren't ready for it, guerrilla-style like the Maccabees I'd been reading about, catching them unsuspecting. It would have to be early in the morning when they were all bundled up on the way to school with a belly full of oatmeal, when they were too warm and satiated to possess an appetite for violence.

That late afternoon I made three snowballs, wet them under a dripping icicle, and hid them in the big chest freezer in our basement. Next morning I told my mother I had an early school project, collected my ice balls, and waited behind the snow berm outside Robert's house. Robert was not athletic and really not a bad guy, but he had been there the day before and he was a good place to start.

In Canada the winters are brutal, so you say your goodbyes inside before going out, then quickly close the door behind. The second I heard Robert's door shut I popped up. We instantly made eye contact. My adrenalin surged and I let my first ice ball fly. Pulling off a lucky shot, I hit him right in the face. With a shriek he went down, a red stain rapidly spreading across the iced stoop.

I ducked and peered over the snow just as I heard his mother open the door. She swooped him up and took him inside. Then I ran. I was elated. It had worked! But he had seen my face. Would he tattle? Would I get that call to the principal's office where my mother would be waiting?

Lucky for me, he kept his mouth shut. He came to school late with a swollen, purplish eye and blood still crusted around one nostril, saying that he slipped and fell. He refused to meet my gaze and never spoke to me again. He also stopped being part of the mob. Encouraged by this win, I stayed in the game and learned to fight, slowly earning respect one by one from the ground up.

It was in the fourth grade that a seemingly small, but tremendously generous, act altered my world forever. A girl in my class who liked me gave me a worn paperback copy of *A Horse and*

*His Boy.* That little girl and C.S. Lewis opened up an amazing portal of literature through which I could slip and go far away. But time was not on my side. I could spend only so many hours locked in the bathroom reading—the only room in the house where I could have peace and quiet enough to allow the story to saturate me. Written words became images, scents, sights, and sounds, enveloping me as if I was sitting there watching the story unfold.

But self-preservation was still my primary M.O. Getting home was always a daily ordeal, where I had to keep my eyes on my back and not dawdle with a book. My half dozen shortcuts bought me escape from time to time, but even they were unpredictable, and I could only run and hide so much. Fights never ceased to be a regular occurrence, but backing down was never an option given to me, and I was bolstered by my occasional taste of victory. There's an old adage that says, "The best defense is a good offense," and I was getting better at both.

One of my best ruses was used against much bigger guys. I'd run away from them, prompting them to give chase. I'd let them catch up to me, then pick up the pace and suddenly drop to the ground. Time and again one big kid after another fell for it: they'd trip over me and hit the ground so fast they barely knew what happened. In that instant I would be back on my feet and on top of them, attacking with kicks, punches, elbows, and knees. I didn't have style or training, but I had fury like no one else. Sure, winning was rare, but I never made it easy for my opponents.

Sticking to my mission of self-preservation, I was often a little monster, spurned on by anger and fed by a sense of retribution. One day during the lunch hour when a bunch of us were playing soccer in our concrete yard, the ball got kicked out of bounds and I ran after it. It rolled right to my classmate's brother, and when I told him to give me back the ball, he threw it over my head. Being the one who chased the ball down, it was *my* ball, and I instantly exploded.

He may have been a grade below me, but he was only three weeks younger and we stood eye to eye. I gave him a solid upper-cut to the stomach. When he doubled over I kicked him in the face hard enough to knock him backwards to the ground. He hit the back of his head and had blurred vision for the rest of the day. I felt bad about it afterward. I hadn't planned on going off on him like that, but he should have given me the ball. (The next time he would.) For the rest of the day, yet again, I sat on that oh so familiar hard stone floor outside the principal's office, staring out the front doors, waiting to be let go.

But I wasn't always bad, even when I was being a little bad-ass. I found that I did care about underdogs like me, and when given the chance, I'd come to their rescue.

A few houses down the street from us were the Aaldenbergs, a family who moved in when I was in the fourth grade. The mister was an ambassador from South Africa, and along with his wife, they had a teenage daughter Alicia and a young son Egbert, who was a few years younger than me and acted as if he'd always been treated like a little adult. Blond, blue-eyed, and frail, with an accent as slight as he was, he desperately wanted to be like the robust hero he saw in his father. Often he would be dressed in military-type shorts and a matching tunic, looking like a miniature Lawrence of Arabia atop his five-speed "chopper" bicycle.

Egbert's bike was, to be sure, the neatest anyone had ever seen, and probably not the best choice for our neighborhood. When a strange kid rode past me on it one day as I was walking home from school, I immediately knew something was wrong. I ran down that kid and yanked him off the bike by the scruff of his collar and proceeded, in today's vernacular, to "ground and pound." Sure enough, that singular bike was Egbert's. When I returned it to his house, his mother came to the door hysterical, looking frightful under runny mascara. She was elated to get Egbert's bike back and told me he was upstairs in bed recovering. I pictured him with a steak over his eye.

It was only a few weeks after returning Egbert's bike that I came across him by the big willow, already knocked to the ground and being beaten up by three young punks. They never saw me coming. Though I didn't have a game plan, I was used to fighting more than one at a time. Seeing little Egbert on his back and those three bastards trying to steal his bike yet again pushed all my buttons of indignation, and my sense of righteous wrath went supernova.

I went after the biggest and most aggressive kid first, hitting him from behind in a brutal tackle that ground his face and chest into the dirt, then I quickly rolled off him and sprang back to my feet. Stunned number two, the second biggest, took a standing clothesline from me like a deer caught in the headlights and went down hard on his back, knocking his wind out. Number three had seen enough, turned yellow, and started to run. But I tripped him from behind and wasted no time pounding his face into the grass.

Then it was back to the first two. The second one was already running the other way fast, but the big kid still had his eyes on the prize. I had never seen him before, and he was more than big and tough; he was persistent. Egbert was now on his feet and attempting to recover his bike, but the bully had his hands on it. In a blur of fists I laid into him as he was pulling it away from Egbert. I caught him with a good left jab, knocked him down, and bloodied him good. That was the last time anyone tried to steal Egbert's bike.

The following year, shortly after the start of the fifth grade, another Jewish kid came to our school. He was about my age, with dark, wavy hair against alabaster skin and piercing aquamarine eyes. He was probably the best-looking kid in the school. He was also in the "special class," meaning he had a distinct mental disability. I had spent a lifetime teaching all of "them" what a Jew was, and here comes this new kid, threatening to destroy my carefully wrought example.

My biggest struggle with my speech impediment was not to be treated like I was mentally handicapped, and I felt like this kid was blowing it for me. Why couldn't he just be normal? I felt embarrassed by him and I didn't know why. But it really bothered me, and that frustration turned to anger.

One Friday, I was walking through the halls looking for this kid before the end of the school day. I went from classroom to classroom, standing on tippy-toes so I could peer through the window of each door. Full of pent-up rage, I was determined to find that kid and teach him a lesson. When the bell rang and everyone poured out into the hallways, I would exact my revenge amid the melee of "school's out."

It was there in that hallway just before school ended when I suddenly came to my senses. Like a bolt out of the blue, I realized just how fucked up what I was doing was—looking to beat up some mentally disabled kid. This wasn't about him; it wasn't his fault. It was about me and the chip I had carefully crafted to rest upon my shoulders. In that instant I realized I was about to become the thing I hated most: the bully. Ashamed of myself like I had never felt before, I swore then and there that I would never fight again.

What I didn't realize is that I had fought my way into extinction and that nobody wanted to fight me anyway. It just wasn't fun anymore for anyone but me. When I got beaten up, I would be outside that kid's front door the very next morning. And if I got beaten up again, I'd be right outside his house the morning after that. A tough guy isn't someone who doesn't get his ass kicked; a tough guy is someone who gets his ass kicked and then gets back up and finishes the job. Because these kids couldn't stop me, or hurt me past my own anger, they left me alone. Not even the six-graders would so much as give me a reason to fight. I had made myself such an untasty morsel that no one wanted to take a nibble.

It wasn't until Halloween of that year, fifth grade, that I was prodded out of early retirement to test the waters of my forced respect. It's one thing to smash someone's pumpkin (though it's not nice), but a cast-iron dachshund boot-scraper was stolen from in front of our house, and my mother liked that old thing. That next day at school, I let it be known if that boot-scraper wasn't returned, I was going to start fighting my way from the bottom to the top, and no one would be spared. I had a feeling it was one of the older Leduc brothers, the one in the seventh grade and the worst of a bad lot. To reach him, I knew I had to put the fear of my retribution into the entire school body.

I made my speech during first recess—still the smallest in my class, still working through a severe speech impediment—threatening each and every one of them. Only Bobby Jones, the poorest kid in the whole school and my only friend, was exempt. My mother could never explain how, but the very next morning that boot-scraper was back outside our front door. That's respect earned the hard way.

All through elementary school, I had simply wanted assured peace, and I was only afforded such luxury after becoming the most unlikely hippo in my croc pond. I didn't have to fight again until I was in the Army.

# FOUR

## Irregular Army Issue

"Be assured that although the training your son will receive here in the Field Artillery Training Center will be physically and mentally demanding, we go to great lengths to ensure his personal safety and well-being, and to preserve his human dignity."
—Henry M. Hagwood Jr., Colonel, Ft. Sill, OK, January 21, 1984

*This was the first time my mother received in writing, from a government official, the promise that the personal safety, well-being, and human dignity of one of her children would be ensured, even at great lengths. She was not assured.*

I joined the Army at the advanced age of twenty due to sheer desperation amid the humbling admonishment that I had failed in the real world. I was out of options; it was either the military or, more likely, jail. Getting kicked out of my cramped third-floor attic apartment and then being fired from my dead-end job at the Smorgasbord in the same week was only the start. Getting caught up with the wrong guys and almost getting my head blown off trying to rob a gas station was the end. In an inane act of self-salvation, I leapt once again from the frying pan into the fire and joined the Army.

Everyone who knew me, my family doubly, shook their heads in disbelief. I'd always had an issue with authority. When I was ten years old, my teacher wrote in my permanent record that I was "self-reliant, antisocial, and uncooperative." That was also the year my mother remarried. I gained a father and a brother my age, Harvey, and five years later our family moved from the small rural pond of Ottawa to the vast urban ocean of Chicago for my father's work. Harvey lived with his mother in Toronto and would visit us on weekends, but when we made the move to the States he came with us. I went from a small-town school of barely three hundred to the immense Oak Park-River Forest High (OP-RF) with an enrollment of over three thousand.

Talk about culture shock. Back home our Nepean gym was the cafeteria, divided in half for boys at one end and girls at the other. Our soccer field had no bleachers; at games, everybody stood. In contrast, OP-RF had a field house with a basketball court in the center surrounded by a track, a weight room, two Olympic-sized swimming pools, eight tennis courts, two baseball fields, and two football fields, one with a five-thousand-capacity stadium.

But this wasn't the only culture shock. The human diversity was fascinating too.

There were a whole lot of people a whole lot darker than me and I didn't stick out anymore. We quickly discovered, though, that they had a culture all their own that was completely foreign to us. Our "elephant" pants might as well have been two burlap sacks in contrast to their ultra-hip straight-leg jeans.

One day, my sister Esther pulled the pick out of the back of a girl's "fro" and tapped her on the shoulder to hand it back to her, thinking the girl had accidentally forgotten it. It was innocent enough but not received well. So much for our Canadian politeness. I also realized that American Jews were "white," which really blew my mind because that meant I was now white in their eyes, and that deeply contrasted with my sense of self-identity as part of a minority.

While it had been a certain kind of hell to be the only Jew in my school in Canada, "White America" did not appeal to me. My high school, steeped in duality, held two senior proms: the "official" one held at the school, and the Black prom held at a banquet hall. More comfortable with the outlier crowd, my brother and I chose to go to the latter, and trust me, that's not a party you can crash. You have to know more than a few people and you still have to get special permission. To the guys who knew my brother and me, we were "the Canadian Jews"; by virtue of that ethnicity we were allowed in. The fact that I didn't have any good shoes and had to wear my work boots with my "good" pants was kindly overlooked.

High school became marked by clashes with teachers and the resulting suspensions, and I went from being a bookworm to selling pot with my brother. I was the muscle and he was the mouth, and I started to lose my way. One day, instead of being in school, we pulled up to the projects just outside Cabrini Green in our '54 Rambler to score some pot. For five dollars you got a tiny Manila bag filled with "dirt" weed, all orchestrated through the mail slot in the door of a second-floor apartment. It wasn't until we were back in the car that I realized they had given me, the White Boy, Cheerios instead of dope.

After taking some serious shit from the two guys in the back-seat, my buddy started to turn the key to drive away. "It's not worth it" he was saying when I stopped him.

"If they shoot me, a white boy," I reasoned, "then the cops are going to come out, and the five dollars isn't worth the hassle."

Back up those steps, choked by the heart in my throat, I put the envelope back into the slot and knocked hard. Sure enough, another packet appeared. I grabbed it and flew down the stairs to the car idling outside. In a place where people got shot all the time, I risked it for five bucks. If nothing else, that tells you I was not thinking straight.

I barely made it into a state school, W.I.U., aka Where Idiots Unite. I didn't know anyone there, and without family or friends I felt isolated and alone—until I found rugby. I had never played an organized sport, but a guy on my dorm floor was on the team and I was intrigued. Without any other clear direction, rugby became my religion, and the team became my family. But you can't major in rugby and drinking beer, and going to the bar at noon to watch *Green Acres* and drink cheap two-for-one screwdrivers is a short road to nowhere. I got kicked out of college at the end of sophomore year, and not just for bad grades but for bad behavior as well.

Shamed and ashamed, I ended up back in Oak Park. Lacking motivation, my mother had to help me find and pay for an apartment, and I started a series of dead-end restaurant jobs while attending a local community college. The only time I felt like somebody was when I stepped onto that rugby pitch; otherwise, I was a loser going nowhere fast.

And then the shit hit the fan.

After I lost my job bar-backing at a Swedish smorgasbord place, I found a job in River Forest as a dishwasher in a restaurant owned by Tony "Big Tuna" Accardo. Twenty years old and riding my bike to a dishwashing job, while everyone else I knew was moving on and up with their lives, really sucked.

One of my buddies on the rugby team drove a big old Cadillac and ran guns for the mafia in the summer to pay for his winter expenses. He had given me a lift a few times, and after explaining what he did, suggested I join him. The pull was strong, so I did. It wasn't long before I wound up at the gas station looking down the business end of a sawed-off shotgun.

Long story short, if you've never hugged the ground in the middle of a field as the searchlights of four police cars circled, or you've never had to streak across eight lanes of freeway at night, or had to crawl through scrub for miles to evade the authorities— you've lived a sheltered life.

Soon after that close call, over two hundred American Marines were killed in Beirut, Lebanon, and I thought there was going to be a war. The deaths of those men really bothered me, and I was desperately searching to find my way.

Make no mistake, joining the Army is an act of sheer desperation, a last-ditch attempt to salvage one's life. I was acutely aware of my own failings and the inevitable downward spiral I was on. I also needed a way of cutting the apron strings and ultimately taking responsibility for myself, which I had yet to do.

A week after the bombing, I walked into the Army/Marine recruiting center. All I knew was that the Marines were supposed to be the toughest, and that's what I wanted to be. But the Marine recruiter was out to lunch, while the Army recruiter was eating at his desk. He showed me a Beta videotape of soldiers stringing communication lines through the French countryside, explaining that I would be joining a military occupational specialty known as 13 Bravo. It was only a two-year commitment after which the government would pay for my college and I'd have healthcare for the rest of my life. I thought, *Learn a trade, see the world, and be somebody . . . I'm in!* Then and there I became a two-year man in the US Army.

By the time I enlisted, I'd had my nose broken well over two dozen times, been shot at, and taken a knife in the back. If that wasn't enough, I had almost torn off my left foot, twice, first distally then proximally. Both times the bottom of it was looking back up at me, all the muscles, tendons, and ligaments stripped. I graduated high school walking with a cane and was told the limp was permanent. Running ever again would be doubtful. Yet I'd recovered to the point of being a standout rugby player, one of the roughest sports you can play.

It wasn't until just past the new year of January 1984 that I finally received my travel orders. The Army was not what my folks had planned for me, so they refused to see me off, leaving my sister, Esther, to drive me to the train station. Once I got to O'Hare and checked in at my gate, I was approached by an Army rep. Apparently my time at Triton community college had parlayed into a job promotion, meaning I was to be a PFC, Private First Class, and as such was required to be the mother hen for fifty-six other recruits on their way to Ft. Sill. That's also the moment I found out that 13B was a group of cannon crewman, not so affectionately known as "gun bunnies."

———

I quickly learned that Basic Combat Training was a whole lot tougher than I ever imagined, tougher than anyone imagines. It'll make you wish-you-had-never-signed-on-the-dotted-line kind of tough. Fortunately, I come from a long line of rabbis and horse thieves, and neither is easily discouraged.

Everyone gets pushed to their physical and mental limits and beyond by the drill sergeants. For me, though, it wasn't so much the physical demands as the psychological servitude. I plain didn't like taking orders. Case in point: I was the first recruit named as a squad leader, and the first to be stripped of that title. I also garnered several citations for misconduct. My relationship with the drill sergeant was strained to say the least. The problem was, he just couldn't intimidate me, and that set a bad example for the rest. Call it a personality conflict.

After two and a half months of mutual antagonism and dislike, our relationship came to a boil two weeks before graduation. Drill Sergeant Felder was fucking with me, trying to teach me humility by forcing me to do repetitive sets of pushups. There was a rule that they could only make us do twenty at a time, so he was making me do twenty after twenty after twenty—and I got pissed. When I was finally allowed up, my schoolyard temper flared. As I dashed back into formation, I stomped hard on one of his perfectly spit-shined boots. That's as disrespectful as it gets, and it was clearly done on purpose.

Had I swung at my drill sergeant, it would have made his day. He could have hit me back, knocked me out, and kicked me out all in one fell swoop. But there was no disciplinary action over a badly scuffed jump boot, though it would require hours of careful polishing in tight little circles to regain its glassy finish. Kicking me out was his best remediation.

———

Late that night, the drill sergeant marched into the barracks after lights out. He slammed the door on the way in and flicked on the lights.

"Attention!" he ordered.

We all scrambled in our underwear to the front of our cots. He had been drinking and thinking while spit-shining and he was furious. As we stood blinking in the light and shivering in the cold, he strode over to me.

"This is your last night here, Harris!" He said my name like it was a curse word. "You're unfit and I'm not going to recycle you just to become someone else's problem. Oh no, I'm going to see that you're kicked the fuck out!"

I stared straight ahead without a flinch.

He leaned toward my face. "You hear me, Harris? I'm going to see you gone! You're a Chapter 11, Harris, Failure to Adapt."

I remained still, though my blood was boiling.

"You're gone, Harris," he promised as he stepped back. "Don't even bother to write your folks. You'll be home before the letter gets there."

He spun around and spouted "Now turn off these lights and get to sleep!" then slammed the door even harder than the first time.

I was terrified, not of the drill sergeant, but that I would once again be a failure, that not even the Army would take me. Almost one out of every four of us had been sent home packing, and I refused to be one of those rejects. I needed to finish what I started. I had to accomplish something to stop being the loser I was. I had burned so many bridges in the real world, I was out of options and had nowhere else to go. Basic Training was almost over. All I needed to do to make it through with a passing grade was keep my head down and not call my drill sergeant's bluff.

The last week, War Week, was dominated by tests on everything we had ever been taught during those thirteen weeks of hell. In other words, every technique we'd learned to kill other human beings, in various mortifying ways, all performed and graded while under constant stress and completely exhausted. And that's when something strange happened.

I got good, very good. I received perfect scores on test after test. As each day went by, the whispers grew louder. Only a person with a flawless score could be considered for Most Distinguished Graduate, and when it did happen it was a rare event.

On the last day of testing, my final trial was to fire off an M-109 howitzer, solo. So far, I'd been flawless. Everyone watching me knew what was on the line. As I packed the charges into the tube in the correct order, I glanced at the scoring drill sergeant. He looked as nervous as I felt as a bead of sweat slid down the side of his face.

With the intense flow of adrenalin, I almost closed the breach too soon. Luckily I caught myself, gave the shout of "I see red!" then slammed it shut. To complete the drill, I had to pull the lanyard and hear the click of the firing mechanism. But the lanyard can be fickle, and if not pulled at just the right angle, the mechanism would jam solid. I torqued my whole body as I yanked on that rope, leaving nothing to chance, and it clicked.

My drill sergeant's face betrayed his bafflement—and his disdain. He was miffed at being proven wrong, and he hadn't changed his mind about wanting to send me home. Instead, after scoring 100 percent on all of my end-of-cycle tests, I was named Most Distinguished Graduate of my entire training battalion. In an unheard-of display, every drill sergeant lined up to shake my hand. A few didn't hide their reluctance, my own drill sergeant included, but they weren't given a choice by the senior drill instructor, or by me.

That was one of the proudest moments of my young life. I wasn't used to excelling, and it felt like absolution. They even presented me with a trophy (which I thought was an odd takeaway for a military achievement). They also sent a congratulatory letter home to my folks, along with issuing a press release to my hometown's local paper with a picture of me shaking the commandant's hand.

Awarded the highest honor that the Army could bestow upon a recruit, I believed I had finally found my place in the world. My platoon nicknamed me Superman—not bad for a short Jewish kid from a Canadian dairy farm.

On the day of graduation, we were given our travel papers. The Cold War was in full effect and troops were massing on both sides of the border. Every one of us was sent overseas.

Setting off to my first deployment, in Hanau, Germany, I was full of optimism for a career I hoped would give me a respectable future. I had a uniform with rank, Private First Class, a duffle bag with everything I owned stuffed inside, and my letter of commendation in my pocket. The world was my oyster, and I was going to be Uncle Sam's mother shucker.

DEPARTMENT OF THE ARMY

HEADQUARTERS, US ARMY FIELD ARTILLERY TRAINING CENTER

FORT SILL, OK 73503

24 April 1984

SUBJECT: Letter of Commendation

PFC Adam L. Harris

Battery B, 2d Cannon Training Battalion, USAFATC

Fort Sill, Oklahoma 73503

1.  I commend you for having been selected as the outstanding soldier of your unit while assigned to the Field Artillery Training Center, Fort Sill, Oklahoma. This is a distinction you have earned in competition with all soldiers in your training battery.

2.  You were selected as most Distinguished Graduate in recognition of your academic and military accomplishments during this period. Your achievement resulted from an energetic application of sound judgment and your newly acquired technical knowledge.

3.  Your attention to duty, personal appearance, cooperative spirit, military bearing and military courtesy were exemplary.

4.  You were of direct assistance to the cadre of your training battery in the performance of their mission, that of training you and your fellow soldiers.

On behalf of the cadre of the Field Artillery Training Center, I extend congratulations on the fine achievement and best wishes for continued success in your military service. I am proud to have had you as a member of this command. You are a credit to the United States Army and to this country.

A copy of this correspondence will be placed in your Official Military Personnel File.

Henry M. Hagwood, Jr.
Colonel, Field Artillery
Commanding

*Dear Hairy Harvey Harris,*

*What's up Bro? Coming to you from my new home away from home—Hanau, Germany. I'm stationed with Headquarters, the brains behind the armored division's brawn. We live in an old brick building in an old brick-walled compound that used to be a German Army Post. They call them kasernes around here, and ours, Hutier, is one of a handful of former German garrisons that are now US Posts in this little town.*

*Anyways, it was my new roommate's birthday last night. He's the driver for a Major, and I got my first taste of cheap German whiskey and coke. No ice. You know what they say—"when in Rome"—and I certainly didn't want to refuse the hospitality. Hell, when have you ever known me to pass up a drink? But sugar, caffeine, and booze are a crazy combo. We were up drinking the whole night and I'm so fucking hungover right now.*

*My head is pounding like a whole percussion section. I feel like a small animal that's been dashed against the rocks. The outside looks just a little ruffled but the insides are totally messed up. Thank God there's a cure for this! Seventeen-hundred (5 o'clock to you) is when they let us start drinking again. And, by the looks of it, everybody has this same affliction.*

*It's a good thing we're on different continents 'cause they taught me how to drive a big ol' 5-ton cargo truck. I'm still having difficulty steering it straight as I go down the road. The whole thing tends to shimmy and shake from side to side as I sit in the driver's seat. I got my license and got into an accident right afterwards. Not my fault, I swear! But maybe that's why the folks wouldn't even let me park the car in the garage, let alone drive it. I went into a chain-link fence. But I'm telling you, it jumped right out at me! Good thing my hillbilly sergeant was with me. He laughed like a crazy man. I mean a bit too much, like he was becoming unhinged. I'm just glad he didn't give a damn. I guess we needed something to do the next day.*

*We had to "volunteer" last Saturday for a six-mile fun run. What a fucking oxymoron! It impresses the Sgt. Major to see a good turnout, so our platoon Sgt. Johnson makes us do it. For us it's either run or break big rocks into smaller rocks with a sledge hammer the next day. What kind of shit is that? And if we come in over 50 minutes, then it's back to breaking rocks. I finished in 42 minutes. No great feat but it kept me away from the rock pile.*

*Big Mike wasn't so lucky, or so fast, and spent his Sunday gathering blisters from the handle of the sledge hammer. That poor sad sack walks around with his eyes down, doing the shuffle of the broken and soulless, like his head is too heavy for his neck. At least he won't miss any money lying on the ground.*

*So much for my plans of touring Europe. Seems like this is a 24/7 job year round. Not quite the picnic I was promised, but I can buy Shlitz for 21 cents a can. Can you beat that!? "Beer, beer, beer, said the Privates, lowly men are we."*

*I'm starting to realize that Basic may have been the best part of the Army. Forget about me stringing communication lines across the French wine country like I was promised by the recruiter. Learn a trade and see the world, my ass. I'm not even using my 13-B field artillery training. For now I'm just a grease monkey, but I do know how to change a truck tire like a pro.*

*So that's the long and the short of it, and on that happy note I'll wrap this up. Take it light and make it right! 'Cause that's the only way to go when you're across the Big Pond and it's too damn far to swim home.*

*Love, your brother in green (and that's an order),*

*Adam.*

# FIVE

## The Lie Detector

For anyone's first military deployment, there is a certain amount of concern, if not downright fear, especially if your first post isn't even on American soil. When you're the last person to know where you're going, and you have no say in it, uncertainty abounds. But when, as a newbie, you catch your boss getting stoned on gasoline, you know you're definitely in for a strange ride.

Squad Sgt. Hillbilly, my direct superior, looked like a squirrel graduated to human form without the intellectual upgrade to go along with it. He was small, wiry, and nervous, and viewed the world with contempt through oversized eyes. He was also Platoon Sgt. Mean Joe Johnson's whip. Calling Hillbilly the perfect henchman gave him more credit than was due. He was nothing more than a little dog with a big bark made even louder by the boldness of his master. In other words, if we balked, he would threaten to run and tell Mean Joe.

A few days in, as the new guy, I was graced with toting our tools back to the shed while the rest of the guys went to wash up. Coming up on our mattock shed from behind, I saw Sgt. Hillbilly. He was sitting on the cold concrete floor with his legs splayed out in front of him, leaning over a metal bucket with a dirty towel over his head.

As I rounded the corner, he blurted out, "Spades!" He then scrambled up and gestured out the doorless opening at an airplane passing overhead. The towel fell to his feet.

I looked at the bucket and fallen cloth, trying not to get caught up in his craziness. "What?" I said.

"Spades . . . everything looks like a spade."

"Like a shovel?"

"No," he snapped.

I took another guess. "Like a playing card?"

A wide grin spread over his face. "Yeah," he said, suddenly contented.

"Ahh," I said with a nod, as if I got it.

"When I huff gas fumes, it makes me feel like I'm home," he said. It also apparently gave him bizarre hallucinations, which he seemed to thoroughly enjoy.

The next week was Mary's birthday (Mary being the last name of one of the guys—remember, it's the military), and we were in the field. It was my first time on maneuvers with the "real" Army. Mary and I were built like bookends: we stood eye to eye, had matching racks for shoulders, and kept quiet dispositions that belied the potential for mayhem. Only a fool wouldn't see the sparkle in Mary's eye that let you know his still waters ran deep.

Sgt. Hillbilly had picked up a couple cases of German beer using the event as an excuse to imbibe. No cake or candles, but in the field you rough it—and of course we weren't supposed to be drinking, which is why the party was held in the rear of a truck.

Just after sunset, the lot of us climbed into the back of the five-ton where Sgt. Hillbilly had stashed the beer for a quiet little birthday celebration. As he explained, this was to help improve squad morale. We all smelled bullshit, but when someone says "free beer," privates stop asking questions.

It's hard to gauge the passage of time when you're in the back of a dank, dark truck drinking dank, dark beer. But by the time everyone else was passed out, Sgt. Hillbilly had a great idea: why not take a drunken joyride in the middle of the night in a truck bigger than a small house . . . hell, in a truck that could carry a small house!

Two of the guys had roused themselves and shambled off to their cots, but Mary and a guy we called NewYork were still passed out in the back. Hillbilly climbed into the driver's side, I hopped into the passenger side, and he started it up.

*I guess this is what guys in the backwoods do for fun*, I thought as we drove off.

We were off-roading without any lights so we wouldn't get in trouble. (You know you're inebriated when that makes sense.) I have never been terribly impressed with roller coasters, but that was one ride worth the price of admission. Anything that stood in our way was fender fodder, and small stands of trees were the sergeant's favorite target.

At some point we realized it was getting close to dawn and therefore time for us to get back. As a grand finale, Hillbilly showed me how to do figure-eights in a potato field. The rear wheels slid out to the side the whole time as if to catch and pass us up until we straightened and smashed right through the guardrail to get back onto the road.

Back at our campsite, we parked as quietly as possible. Hillbilly and I helped Mary and NewYork out of the back, where both had slept soundly throughout the lurching ride, then got into our cots just before the privates who pulled the last fire guard duty yelled "wake up" as loud as possible.

Right after morning formation, Sgt. Johnson grabbed me hard by the shoulder. He tried to yank me but I just stood there, dumbfounded and still mostly drunk.

"C'mon, the Top wants to see you." He tethered himself to my shoulder as he led me in the direction of command central.

There was an uncomfortable silence inside the top sergeant's tent. He gestured with his eyes to a chair in the middle and Sgt. Johnson pushed me toward it.

"Sit," said Johnson, just shy of a shout.

In front of the folding metal chair was a small table with a truck battery on it. The commanding major was there as well, and both he and Top were red-faced, as if they had been yelling and some unfortunate guy before me had been the target.

I sat and Top leaned his face right next to my ear, with the major hovering over him. "Have you ever taken a lie detector test before?"

I was aware of the power these three had over me, but I didn't see a lie detector machine anywhere. I glanced at the battery and saw cables under the table. I began to sweat.

"No," I said.

"Do you know who drove the five-ton last night?" Top thundered. His face was to my left, Sgt. Johnson to my immediate right, and the major staring at me dead on.

"What?" I replied as innocently as possible. *I'm such a bad liar.*
He asked again, louder this time.

"No," I said again. *I'm lying, and I know they know it.*

"Let me light him up." Johnson stepped over to the table and picked up the cables.

"This is how it's done in the field," said Top, "and believe me, you're going to squeal like a rat with its nuts in a trap."

I could see the major sneering as Sgt. Johnson walked toward me with a clamp in each hand, the other ends of the cable hooked up to the battery.

I swallowed hard in terror.

"We know you drove the truck last night," yelled the top sergeant. "Fess up!"

Sgt. Johnson touched the two clamps together. Sparks flew and I jumped in my seat. *Was this really going to happen?*

With Sgt. Johnson menacing me with the jumper cables, I told them we had a few beers to celebrate Mary's birthday in the back of the truck and I had passed out. I reasoned that because the other guys had passed out and didn't remember anything, they weren't in the hot seat. Since I was the passenger, why couldn't the same alibi work for me? But Top kept insisting that I had been driving, almost like someone had said so.

It then occurred to me that Sgt. Hillbilly was nowhere around.

My mind started racing. Somebody might have heard the truck, then saw me getting into my cot late and ratted me out. Or maybe that country bumpkin thought he could label me a troublemaker and feed me to the pigs like slop. Or maybe they just thought the new guy would break the easiest. But they—the major, top sergeant, platoon sergeant, and squad sergeant—all got it wrong.

I'd grown up stained with a school-ground code of honor, and I wasn't going to squeal no matter what.

After an exasperating half hour of threatening me without any results, Top finally said, "Well, how do you explain this?"

He led me outside with the other two in tow. There stood our five-ton, looking not too much worse the wear, save for a few dings on the scratched-up front fender. But the truck did seem to have grown a tail. Caught up on the rear differential was a hundred and fifty feet of heavy duty steel cable with large wooden posts that had been planted in the ground just the night before. We were so drunk we never noticed dragging it with us, even after we parked the truck and helped the guys out of the back.

My smile might have betrayed me but my tongue wouldn't, and all they could do was scowl as I stifled my laughter.

*May 18, 1984*

*Hi Harvey,*

*How's life in the big city? Over here it's eternally cloudy and overcast from morning to night—and always drizzling. You can't ever tell what time of day it is because the sky is always cold and grey, just like the people. Being stuck in Germany is so depressing that even the sky cries.*

*Here at headquarters everybody has a security clearance. It's essential for us to do our jobs, and all you need to get that clearance is to be an American. So, as a Canadian, I'm the odd man out, and it's like this is the first time this has ever happened. Seems to have been a clerical error sending me here, because I can't do anything beyond the most menial labor at HQ without that stupid fucking clearance.*

*And I can't get the clearance without citizenship. But I can't get American citizenship unless I'm on US soil—and they're going to keep me here in Germany for my entire tour of duty. I know 'cause I asked. So I'm perpetually stuck at the bottom of this shit pile . . . talk about the perfect catch-22. What kind of future does this give me?*

*Here's the icing on the shit cake: I received a counseling statement from my hillbilly squad sergeant and signed by his boss, asshole Joe Johnson—our platoon sergeant. These guys have been riding me from day one. I keep my quarters ship-shape just like we were taught in Basic, which was crazy strict. Still, Sgt. Johnson wrote in a progress review that I "seem to be inherently lazy and dirty." My roommate's side is by far messier than mine and I can't help but wonder if it's an antisemitic slur. I've been called a dirty Jew before.*

*Sgt. Johnson couldn't charge me, but threatened me with "Insuboordination And Clothing, Equipement and other Military Proporty Unclean" as well. Do you believe this shit? Most of it sounds as made up as his spelling. So, what do you do and where do you go when the people in charge of you are fucked in the head?*

*At first I thought the sergeants were just picking on me. Turns out 6 people have gone A.W.O.L. from this unit alone just this past year—that's got to be some sort of record. My roommate had just come back from being AWOL when I got here, busted from a specialist back down to a PFC (E-4 to E-3). There's a lot of crazy shit that goes down here that we'd never do back home, but it's like here . . . we don't give a damn.*

*My new Sunday ritual: my roommate and I hang out at the local park and get drunk on cheap wine. It's barely passable as vinegar—I think it's the same stuff the Roman soldiers drank. We force each other to drink till we pass out in the grass. It's only embarrassing if you're not drunk. Never thought I'd be a wino, but I guess this is part of "be all you can be."*

*Our days are filled with repetitive mindless tasks—digging holes just to fill them in the next day. And preparing for what we hope never comes. Training to be something that we don't truly want to be. But it's the monotony that's the killer. I can't say I enjoyed one day in Oklahoma, but at least Basic was exciting.*

*The other part of it is we're US government property. We belong to Uncle Sam—so it's like we're puppets stuck in limbo. We have no say over what we do or where we go. And that's not just day by day but minute by minute. We signed those rights away. If somebody tells me to do something, I have to say "Okay." And we never know when we'll be ordered into action. But the Grim Reaper is always hanging around. So listen to this fool and keep your ass in school.*

*All right Harvey, enjoy your freedom—you can thank me with a cold beer when I see you in two years. And drop me a line. If a bribe helps then I'll send a check your way after I receive your letter. If you ever want to hear from your brother again, you know what to do—write!*

*Love, Adam.*

# SIX

## Mean Joe

Headquarters is where I got my second nickname. Everyone in the Army gets at least three and you never have a say in any of them. Part of the indoctrination of Basic was getting a new name to go with your military identity. The next came from our first deployment and was usually something superficial: because it was a girl's name Ambrose Mary was simply Mary, Big Mike was quantified, and you didn't need to guess where NewYork was from. After a year or so of being in the same place, once people got to know you, you'd get the moniker that usually stuck for good. At headquarters (HQ for short), they started calling me "Baby Hulk." I didn't mind it, but like with "Superman" in Basic, it never took. For whatever reason, it was just easier to shout "Harris."

The start to my impressive military career was as one quarter of the "Wrecking Crew" assigned to headquarters. Mostly what we broke were five-ton truck tires. Flat-tire repairman was never my career choice; "breaking a tire" involved hacking the rubber tire off the metal wheel using a mutated axe called a mattock. It was insane how we did it. Beyond good hand-eye coordination, you had to have a thorough understanding of how long that axe handle was or you would either chew up the tire and incur the wrath of Staff Sgt. Johnson, or strike the metal core and have the reverb shoot through both arms like angry bees under the skin.

Akin to the "special" class in school, the Wrecking Crew was comprised of military misfits who were only good for the work of morons. My sin was being Canadian, and breaking tires was as good as it got because every other crappy detail was ours as well —from filling in holes to digging them out, from tearing down dilapidated structures to painting others just to keep them standing, and everything in between.

Staff Sgt. Johnson—known more affectionately as Mean Joe— would sit in the shade and sip from his canteen while watching us work. To avoid a comfortable silence, he filled the void with a constant diatribe. A backhanded compliment was the closest he came to ever saying a decent thing.

Mean Joe was my first platoon sergeant out of Basic. With a wild look about him and a penchant for violence, Joe was a bully who had found his place in the service. Everything I know about motivating individuals the wrong way I learned from Joe. My Mum would say, "You catch more flies with honey than vinegar." But there was no honey in Mean Joe. All he knew was piss and vinegar. He was highly proficient in telling people what to do; he just didn't know how to make them want to do it—and fear is a fractured motivator.

Sgt. Johnson had peaked in high school, joined the Army shortly after graduating, and forever saw himself as Big Man on Campus. But he didn't have the leadership qualities necessary to make it past entry-level sergeant, which is why he was stuck with command of the Wrecking Crew. Frustrated that he had gotten as far as he would ever get, Joe would engage in his favorite sport— targeting anyone below the rank of sergeant. He saw it as his supreme duty to correct every aspect of every inferior person he came across—or rather, put people down as a way to boost his own shoddy self-esteem.

Mean Joe would single a person out, making them stand at attention no matter where they were, and read them the riot act for any perceived violation. Since arriving at HQ, Sgt. Johnson had gotten right up in my face daily, belittling me in front of the new battery I was trying to fit into. My first week he made me drop and do twenty pushups in the middle of formation for having dried toothpaste in the corner of my mouth. Our burnt-out commanding top sergeant was resigned to his alcoholism, so as long as the sergeants below him kept the rabble in check, he didn't care what else they did. That gave Joe full rein to throw his weight around—and he didn't need any encouragement.

At the end of my second week at headquarters, one day back from the field, we had a surprise room inspection. I had only been there for twelve days and all I had was the wall locker, cot, desk, and chair that came with the room. Sgt. Johnson took a quick look around and pronounced the place a pig sty.

"It's just like I found it," I responded.

That was the wrong answer.

Joe was like a big dog with a bone—he wasn't letting go until he was done. He reached into his back pocket and pulled out a white glove. I saw the specialist behind him roll his eyes as Mean Joe got down on the floor and rolled onto his back beside my bunk. He reached under and rubbed the rusty old springs. Then, with a flourish like a magician, he held out a single smudged fingertip.

"Tomorrow you'll be cleaning your room from 09:00 to 17:00, or until satisfactory conditions are met!"

He got up and stared at me intently, smiling like a trainer

thrilled at the challenge of breaking yet another animal. Then he turned and strode out.

That afternoon at formation, Sgt. Johnson announced a GI inspection the following Monday morning, entailing everything from weapons to boots to the coils under our cots.

"You can thank Pvt. Harris for this, and I suggest you get started ASAP. I will be thorough!"

That's how Joe wished us a nice weekend.

HQ was my first post and I had gone into it full of optimism. But it was neither the job my recruiter had promised, nor what I had trained for, and I thought the yelling would stop in Basic. I have no aversion to hard work; if you ask nicely I'll move a mountain for you. But treating me like a petulant mule sets my stubbornness into overdrive. Mean Joe knew he was intimidating, but it's all he knew, and to my core I believed then, as I do now, that petty dictators must not be tolerated. To me it was a matter of self-respect: How does a person respect themselves when they put up with such an asshole?

Like a sword suspended over our heads, the ever-present threat to keep us in line was being sent down the road to the gun battery—because that's where people got killed. In spite of Sgt. Johnson's bullying, the real possibility of death was worse and no one wanted to be transferred. So those of us who didn't go AWOL just put up and shut up.

One day, I received a citation from Mean Joe that went like this:

> During our recent troop assessment I realized that your position in this unit is detrimental to you and your military career. You have no certain job status, which seems to provoke an unmilitary attitude in your behalf. You also

> seem to have a keen sense for procrastination in
> your daily duty responsibilities towards
> cleanliness of your room. With continuing this
> attitude you may find yourself in infraction of
> Failure to Obey an Other Order.

That was his way of letting me know that for little to no reason, I could be given my walking papers any time. In other words, toe the line or else.

We spent the following day in the back forty clearing brush. Mean Joe was proud of his nickname and had realized how much it irked me for him to ask me a pointed question and then interrupt me with "What?" over and over again. It delighted him to see the frustration spread across my face every time I tried to answer, never giving me the chance to finish my sentence. He didn't know I had grown up with a severe speech impediment, and he had no idea of my history or the buttons he was pushing, only that forcing me to repeat myself really bothered me. But he was bored, and teasing me entertained him.

By the time we finished our landscaping in late afternoon and were heading back in, all of us were hot and bothered—but I was seething. Johnson was having too much fun fucking with me to stop and he knew I was close to the edge.

"HARRIS," he shouted at the back of my head as we trudged back to the billets.

I ignored him.

"HARRIS!" again rang in my ears with the sting of his contempt. My neck muscles tightened and the hair stood on end.

*Not one more time*, said the little voice in my head. My slow brew had come to a hard boil and I was done.

It was one of the few times in my life I didn't think I stood a fair chance against another man in a fight. But after a lifetime of of being knocked down and beaten up, it wasn't enough to scare me off. All I wanted was to land one good punch. Even if he kicked the shit out of me, which I figured he would, I just wanted a little piece of him.

I abruptly stopped and turned around. Then I uttered the only words I could think of to leave him no choice but to fight me.

"Go. Fuck. Your. Mother."

That's when something happened I could have never predicted.

Mean Joe turned and ran.

I stood there with my chest heaving and fists balled up, with the rest of the crew ready to witness a major throwdown, and there went Jack-Rabbit Johnson bolting like no one had ever seen him run before. We all exchanged dumbfounded glances, stunned into silence. Had Mean Joe seen me as his ultimate nemesis? As the defiant loser already accepting defeat, yet still ready and willing to fight to the end? Or was he actually scared off by the fire and fury in my eyes?

We watched Joe run all the way past our headquarter's billets right to the command building, where the sergeant major held sway over the entire post.

The week before, unbeknownst to any of us, the sergeant major had witnessed a scene from a distance between Mean Joe and me. In the field with headquarters for my first time, we had been setting up tents in an area that had welded metal frames set into concrete pads. All we had to do was pull heavy-duty canvas covers over top of those skeletons. But they were unrelenting and uncooperative, and we were still struggling to set up the first tent, the largest one, as it neared noon. Even with all of us tugging and yanking, the friction was too great for us to slide the wet skin over its frame. Just as Sgt. Johnson called lunch, it occurred to me that we simply had the wrong angle on our pull.

Instead of breaking for lunch with my team, I climbed up the frame and across its top spine, leaning head first toward the crumpled canvas. I grabbed the edge with both hands and started inching my way up the frame ass-backwards—and it worked! As I went, the tent slid right over the frame. By the time everyone else turned around to see, I had the canvas halfway over. The already discouraged Mean Joe took umbrage at the effort, clearly bent that I had shown him up, and berated me for not following orders.

So, when Mean Joe busted into Command with a story about the insubordinate Pvt. Harris instigating a fight, the sergeant major already had his own judgments about both of us. He didn't want to bust me because he had seen me in action and knew I was a hard worker; he also knew what a jackass Johnson was. So even though I was in the wrong and Joe had the stripes, the sergeant major spared me a court martial and instead issued me a "lateral promotion"—which might sound like a reprieve, but turned out to be yet another frying pan into the fire move for me.

That same afternoon, I was relocated down the street to the howitzer battalion where, the sergeant major reasoned, I could at least practice my military occupational specialty: cannon crewman. From changing truck tires to blowing things up, none of it amounted to any real-world application, so it wasn't much of a career move. And being in the midst of a bunch of gun bunnies wasn't my idea of a "promotion," sideways as it was—but at least I didn't have to deal with Sgt. Johnson anymore.

*Hey Harvey,*

*Greetings from the Fatherland. Ah . . . there's nothing like being a Jew in Germany! Deutschland does have its perks—at the Mickey-D's you can get a beer with your horse-meat hamburger. In fact, pretty much every place sells "brau und schnitzel." Schnitzel (that's the sound you make as it comes out of you) is all the unused odds and ends of the pig, from snout to tail, mashed together into a patty, flattened out, breaded and fried. That's why they drink so much beer—you've got to be drunk before you eat something like that. In this little while, I think I've already eaten enough schnitzel to last a lifetime.*

*So now my favorite place to eat is a Korean restaurant—this is one of their menus I'm writing on the back of. It's funny, the Germans hate foreigners and the Koreans like to stick to themselves, so the two get along really well. There's also a lot of Turkish people over here and they try to fit right in, which rubs the Germans the wrong way. So there's skin-head nationalism, neo-nazism, and anti-immigrant (anti-Turkish) sentiment as a backlash. Here the Turks are treated as third-class citizens. They do all the dirty jobs that the Germans don't want to do—kind of like Mexicans in the States.*

*My German's getting better, but it's such a harsh guttural language—all function and no form. Language is definitely a reflection of those that speak it. Whenever something new is invented in Germany, they don't come up with a new word, they just mash a bunch of existing words together. So a bra is a titsenholtsen (no explanation necessary), gloves are called handschuhs (shoes for the hands), and a short person is referred to* as *dreikäsehoch—three cheeses high. And no, that doesn't apply to me . . . but it does to you!*

*We like to have fun with the language. Instead of saying danke schon for thank you we say "donkey shit," and to say goodbye we say "o' wiener-stain" for auf wiedersehen. I don't know what's funnier—the locals who know enough English to correct us . . . or the ones who don't. Oh, we can be such ugly Americans sometimes.*

*Did I tell you about what happened on the autobahn the other day? I was driving my five-ton and when we hit a bump in the road, the steering wheel came right off in my hands! I was so freaked out I didn't even think of hitting the brakes, and we crossed over into incoming traffic. The truck swerved like crazy and the guys in the back all started screaming bloody murder. Thank God the guy next to me was a mechanic and he jammed the wheel back on the post. I think I lost a couple years off my life in those few moments.*

*Oh, and get this . . . Big Mike has disappeared. Talk is he went to Miami, his hometown, and we're all jealous. At last Monday morning's formation, there was an empty spot where Mike should have been, and everybody knew he had flown the coop. Yet more evidence that this place has the morale of the gallows!*

*So smile, you could be me! Well Home-Boy, that's my pep talk for another day in an other way.*

*Love, Adam.*

*PS: Are you getting these letters? Do I need to send you my address again?*

# SEVEN

## Gun Bunny

Within twelve hours of being transferred to the gun battery, I managed to piss off every single Death Dealer on post—most notably my new top sergeant.

At HQ, I had a large room all to myself, but here I had to share a small room with three other guys. I had the bright idea that offering up a bottle of Jose Cuervo to ingratiate myself with my new bunkmates was a good plan for my first night. After the four of us finished that bottle in no time flat, one of them said to me, "You know we have a bar on this post."

A bar on post is not that unusual, but for it to be opened to enlisted riffraff like us was a real treat. No one had to ask twice if I wanted to go, and Kashkari—AKA Kash—the ranking specialist, led the way with Roland the Gunner, Big Ben, and me in tow. Access to booze was great, but I wasn't expecting the little honky-tonk to be blaring with Country and Western music (what I called White Man's blues and the Rebel Yell). As I stepped through the doorway, I was immediately transported deep into the Southern Red-Neck Belt. Not quite my cup of tea when you consider that the two most prevalent things I know about the American South are that it was built upon violence and inhumanity, and that I'm not ever going there.

As a Jew there are a whole host of countries I'll never visit, mostly the ones where they'd like to behead me on camera. The southern states fall into that same league. When my dad was doing his medical residency in Houston, Texas, a cross was burned on his front lawn—the calling card of the Klan, just coming by to say hello. Yup, sounds like the perfect place for an uppity Jew like me to get killed. In history class we were taught that Robert E. Lee was an honorable man. But if someone fights for a dishonorable cause then that person is dishonorable, end of story—and there is absolutely no defense for slavery. So you can keep your grits and your biscuits and keep looking away, Dixieland. Old times there are not forgotten.

After a few drinks I was starting to nod off. Beyond the fact that I was over my limit, it had been a seemingly endless day. While my new roommates were still going strong, I excused myself and made for the front door. That's when I saw her.

Her angelic face topped a thirty-six, twenty-four, thirty-six body made for sinning—a brunette to make the hungriest man forget all about his beer and steak dinner. But she was dancing with a "cowboy." (Having grown up on a Canadian dairy farm, I always thought American cowboy boots were only good for killing cockroaches in the corners. Where I come from, real Cowboys wear rubber boots and they know enough to take their hats off when they're inside.) As I sauntered past them, my eyes were riveted on that estrogen super-magnet when I came face to face with her partner. Seems he didn't like the way I was eyeballing his gal, but I really didn't care. I was just trying to remember what she looked like for later.

His shirt's pearl buttons were undone two-thirds of the way down and I asked him in jest, "Don't the rest of those buttons work?" I figured if he got the girl, at least I got the last laugh and continued on out. What I didn't know was there was a back door to the place—and that quip was spark enough for the cowboy to head through it, determined to meet and beat me out back.

My new barracks were behind and off to the side of the bar, so I headed around back to go straight there and off to bed. I was feeling pretty good. I was nicely liquored up, my roommates seemed to be decent guys, and I no longer had to deal with Sgt. Johnson. And then the cowboy stepped out of the shadows.

He was a lot madder than he needed to be, and I was not impressed by his false bravado. Keeping a sense of humor, I suggested that fighting was not the solution. However, if he really did want to fight, I told him exactly how I was going to kick his ass. I didn't mutter, I didn't mumble, I didn't stutter, and I didn't stumble. I simply told him without malice, looking him dead in the eye, giving him the chance to look deep within himself.

When I'm at the boiling point, people have told me it is as if I grow bigger. I was halfway there when the cowboy decided he didn't want to fight after all and turned to walk away. But after only two steps, out popped the brunette. She said something but I was fixated on those full, pouty lips and not the words that came out of them. Clueless to what I was responding to, I said the one word I would have killed to hear come out of her mouth: "Yes."

Her response was to slap me, and hard.

In an instant, I was jumped from behind all at once by what turned out to be six guys. They had no idea I was used to fighting more than one at a time, and I didn't go down as easily as they hoped. For nearly five minutes, the seven of us fought and wrestled like a ball of angry cats from one end of the football field–sized wash area to the other before I realized the guys were MPs. Six against one, especially unannounced, is not much of a fair fight—and that's just how the Military Police like it.

Finally tripped up, the bastards put me in hand and leg irons because I had put up such a struggle. I was unceremoniously thrown into the back of their Chevy commercial utility cargo vehicle, "cuck-vees" as they were called, and carted away to the correctional facilities. The MPs had brilliantly assumed that because I was being struck by a woman, I must have been the troublemaker. They weren't entirely wrong, but in this case the only person doing any assaulting was the brunette bombshell, and she walked away scot-free while the MPs were still busy subduing me. They didn't even get her name, which was a pity because I would have loved to see her again, despite the fact she was the kind that makes a man lose all civility, and who slapped me to boot. But common sense and common courtesy are neither MP requisites nor strong points.

At the Disciplinary Barracks, the commanding officer figured out that the person who had actually gotten hit was the one they had arrested. The verdict was no harm, no foul, and no charges were pressed. No sorries either, of course, just a call to my new top sergeant to come pick me up . . . at three in the morning.

Fighting at the gun battery had become such a problem that the day before I arrived, Top had laid down the law. If anyone else got arrested for "any physical altercation," the entire barracks was going to have to stand at attention outside in their underwear, from the moment he got the call till he got back to the billets with the offending party. That's how I was introduced to the rest of the Death Dealers.

Standing in formation in the cold in their skivvies waiting for me, you can imagine how happy they were to see me. The next morning things were a bit icy, but by then my roommates had a better understanding of what had happened and stood by me, shielding me as if they knew something or someone was coming.

That something was an invitation, and it came from the enforcer, Mac the Mechanic.

Mac was the equivalent of a double-barrel, ten-gauge sawed-off shotgun—short, blunt, and full of firepower. There are a lot of unspoken jobs in the service, and it was Mac's job to find out if I could be trusted or not.

The criminal investigation division (CID) is the Army's secret police. When someone is transferred out of the blue on "questionable circumstances," they are often suspected of being an infiltrator, or agent, for the CID—whose agents were known to be the worst scum to slither in green. A buck-private with eleven years in and nineteen to go, Mac was given this job because he dared to go where others would not. He had done it before and he'd do it again. In fact, he enjoyed the aura it gave him, the notoriety that made other enlisted step aside.

"Harris, you're taking lunch with me," he said, then turned to lead the way. Full of curiosity and trepidation, I followed him off the post. I had never eaten lunch on the Strasse during the workweek, and I was starting to think that this going to be fun—that is, until we turned down an interminable alley. As I nervously watched Mac's hands, I no longer thought about lunch or making small talk.

I was pretty sure he didn't have a gun, but I was almost positive he had a knife. An enlisted can't have a handgun, unless they're a medic, and an M-16 can be a little lengthy in close quarters. So knives, big knives, were prevalent. When he reached into his jacket, my suspicions went into overdrive. I braced for anything that might happen next, as if someone had just whacked the hornet's nest of my adrenal cortex, setting all my muscles and nerves abuzz.

As he pulled his hand out, I wondered if my eyes were deceiving me. In Mac's hand was a little hash pipe. "Wanna get high?" he said.

I wanted to laugh, but I didn't. I also wanted to kiss him, which I didn't do either. Instead, probably too eagerly, I said, "Yeah, I want to get high."

I settled into relief, thinking how I would much rather be bleary-eyed than bleeding any day.

After passing the pipe back and forth a few times, we strolled out of that alley side by side with goofy little smiles, my appetite forgotten and my acceptance gained, both of which were highly unusual.

Mac never asked me to lunch or got me high again, but as far the Death Dealers were concerned, I was now "good to go."

*Dear Harvey,*

*I got your letter. Thanks—it means a lot. Just so you know, I am Uncle Sam insured and you're my beneficiary. Yup, if I get snuffed then you get $35,000 to party with. Never thought I'd amount to that much so fast and all I've got to do is die to cash in! That reminds me of a joke: What did the German guy say to his brother after he pushed his mother off a cliff? "Look Hans . . . no Ma!" Oh, I can hear the groans from here.*

*I'm sure you shook the check out of the envelope first. Don't sweat it. I know you're a starving student, and when you're a big-time movie director you can pay me back.*

*Good news, I got transferred to a howitzer battery to work on my MOS (military occupational specialty). Finally, I get to play with the big guns, and the M-110, or Eight-Inch, is the biggest thing around. There's nothing in the world that can "shoot and scoot" like them.*

*I'm still stuck in this same little town, Hanau, but anything's better than breaking truck tires, and a change of scenery will do me good. All in all, I feel like I dodged a bullet getting the fuck out of HQ. This guy, a PFC like me, died in a horrific truck accident recently. He was driving a 5-ton like I used to, towing a trailer with a big ass M-109 howitzer (they're like the little sisters to our M-110s) for a local town event celebrating German-American partnership.*

*The truck's brake lines to the trailer gave out as he was going down a long steep hill, and the howitzer got away from him. Get this . . . the fucking thing caught up to him and smashed into the truck, killing him. A couple of German police providing escort got real banged up as well, all because the truck he was driving was a piece of shit—just like mine was.*

*And that's not an isolated incident. A private got crushed between his howitzer and the retaining wall of the wash area last month. And a sergeant was steamrolled by his own M-88 (a track vehicle for towing other track vehicles) when it tumbled over on him. So it just goes to show—machines of war will slake their thirst for blood one way or another. But that's just an occupational hazard. I keep my head on a swivel.*

*I'm fitting right in here at my new digs. The François Kaserne is home to the 1st battalion of the 40th field artillery regiment in the 3rd armored division (Third Herd they call us). So now I'm a gun bunny and when we go to the field I get to hump joes. How do ya like that? (A joe is a great big bullet, in case you didn't know.)*

*On this post there's a bunch of batteries, five I think, and all together they make up the 1st battalion. FYI: a battery, or a company, is comprised of over a hundred men, broken down into various platoons. Each platoon is made of squads, and each squad is 8-10 guys—which makes up a gun crew. I'm with Alpha Company. We're the "Death Dealers," and we are all about blowing shit up.*

*I've met just about everybody. The guys here are very unpretentious and my new roommates are really nice. Remember being told that before a speech you should picture your audience in their underwear? Well, it's a long story but I don't have to imagine what anyone here looks like in their skivvies. Been there . . . done that. The top sergeant seems to be a decent fellow. We had a chat during a car ride, and I'm sure we'll get along just fine.*

*However, my new platoon sergeant, Sgt. DumbOx (rhymes with lummox) is a muscle-bound moron of the highest order. He's twice as strong as an ox and half as smart—two-syllable words hurt his head and three syllables make him crazy. He's already threatened me with physical violence a couple times, so I just lob big words at him and watch his brain explode. I've got to be careful not to let him corner me alone. Poking the bear can be a dangerous game . . . but danger is what we do.*

*You should see, above our mess-hall door is a big painted sign with a picture of an arm in shining armor, and its gloved hand holds a battle-axe. There's no body—just the arm and axe against a blood-red background with our motto All for One scrolled underneath. In big black letters on one side is written DEATH DEALERS, and on the other DINING FACILITY. The joke is: Death Dealers is the name of the restaurant and the arm is what's for dinner. I'm pretty damned sure that human flesh tastes better than the mystery meat they're feeding us. My drill sergeant said we taste like salty pork. And stranger still . . . I believe him!*

*Oh, and tell Jerome he was right, the European women don't shave or use deodorant—though neither one bothers me. You should also tell him that they're all big breasted, don't wear bras, and are incredibly sensuous. I don't know if it's a blessing or a curse, but I can't understand a thing they say. I can tell them anything and they just smile, take me by the hand, and lead me home!*

*Check this out, I was in Munich last weekend, and they've got nude beaches and beer gardens. Is there anything better than being drunk and naked in public? I met a stewardess and almost went AWOL. Sorry—no details on this one. She could be your sister-in-law . . . if I ever find her again. I really think I fell in love. We spent the whole weekend at a hotel together. Clean sheets and room service, now that's what I call heaven. She is definitely "One of the Ones" and she felt so right in my arms. But I know it's the wrong time and place in both our lives.*

*Sunday at the station, she was heading for Berlin and I was going the other way to Frankfurt. I got on her train to go with her instead. She tried to talk me out of it, but I didn't want to leave her. At the last second I came to my senses, gave her a quick kiss and ran off just as the doors were closing. I caught my train back to Hanau just fine, but I can't shake that feeling.*

*I can still smell her perfume . . . her hair. Every time I think of her I feel good and bad at the same time. I miss her fingertips. What if she was my one shot at true love?*

*So think of the love lost and have bittersweet dreams—like the kind I have sleeping on this lumpy old mattress. It's off to the field next week and I'll write you from there.*

*Love, Adam.*

# EIGHT

## Hobbled Goblin

What I'm about to share has haunted me for over thirty-five years. I didn't tell anybody about it after it happened, and I waited thirty years to talk about it at all. I always felt embarrassed because of who my assailants were that day, and how easily they took advantage of me. But that's not the whole truth. It's what I did to them in return that still makes me shake my head and say, "That's not me—that's not who I am." But it *was* me and I must take ownership of my actions. I admit that after all these years, it's one of those events that's still on my conscience, and I still feel the sting of shame.

Nothing recedes like success, and no one feels the ravages of time like a person whose defining characteristic has been strength. In my glory days I was surprisingly strong, not just for a person my size, but for anyone of any size, and I reveled in it. It was my equalizer.

The first time I saw a weight room was in high school after my family had moved to the States, and I immediately showed a natural aptitude to the rigors of lifting heavy weights. When I found rugby, it was like a light switch went on and the front row was my calling. That athletic pursuit was the perfect marriage of all my fortes—brawn, endurance, and tenacity.

But it takes much more than determination to play week-in and week-out in a competitive league like the Chicago Area Rugby Football Union, and I was the loose-head Prop for the East End Ruffians for three years. I used to full-squat eight-hundred pounds in the basement of the little local YMCA. (I didn't tell them I was Jewish.) On alternating days I'd do eight-hundred push-ups, fifty at a time while watching TV, and run eight miles a day, the last one backwards. All for eighty minutes of the greatest game I ever played. That pitch becomes your universe, and when you own the ball you are the supreme being. During my last year with the Ruffians, before joining the Army, I'd have to go to work after the game. Beat up and bruised, I'd get on my bicycle and ride to my dishwashing job at the Mafioso restaurant. It was with that physical conditioning and blue-collar mindset that I joined the Army—which in part explains how I became the Most Distinguished Graduate of my field artillery training battalion.

———

After my right foot was crushed by the road wheel and I was re-cuperating in the "rear," my physical therapy was left up to me. Right across the Kaserne was a small bar tended by one of the most naturally beautiful women I have ever seen. Hilde looked like Brooke Shields' far more attractive sister, and Brooke is cer-tainly no slouch herself. Guys would nurse their beers just to spend time looking at her. I would order a double-strength tequila sunrise and a "dessert," my very first concocted drink: three shots done in rapid succession of Escorial, ouzo, and amaretto. The fiery shot of Escorial was like Aqua Velva mixed with kerosene, followed by a strong, cheap ouzo, and then smoothed out by the sugary amaretto. Then it was off to the dances on my crutches.

In the military, when someone has an injury requiring crutches, that person is immediately labeled "Gimp." Generally, the first person you know who sees you with crutches offers up the remark as if it were witty, like they just made it up. The only proper response was a sudden solid whack with one of those crutches, without mercy—even if they were your friend. The second person to see you and call you Gimp was also going to get hit, and even harder. The third person was going to think better before trying to rename you. At least that's how it worked with me.

It was on those damn crutches, on an autobahn overpass making my way to the dance hall late one afternoon, that I was overtaken by two lovely young women. Actually, one was cute and petite while the other was rather big and scary, but I'm not picky and I don't like to pigeonhole the ladies. So we started talking and they began flirting with me. All of a sudden, the petite blond one, with a mischievous grin, lifted up her crop top and flashed me her perky little breasts—exquisite creamy white mounds with raspberry-red nipples. My eyes got big as saucers, and that's when the big brunette next to me reached over and deftly plucked my money and ID from my front shirt pocket.

As if rehearsed, they broke off from either side of me and ran ahead ten yards before stopping to turn and taunt me. It happened so fast, I was stunned. Just as quickly, I felt monumentally stupid for being duped so easily.

Before making their getaway they paused to laugh at me, the crippled GI, just out of arm's reach. Then they turned to run. I knew I couldn't let that happen; there was no way I could catch up to them, and I'd never see my military ID again if they took off, which would get me into very hot water. They also had all of my cash. The idea of no more booze for the rest of that evening was intolerable.

Being raised with sisters has several distinct advantages. I learned to look at women as people first and as the opposite sex second. I also learned how to push buttons. Men might be stronger but women are smarter, and the less a man agrees with that statement the more they've had the wool pulled over their eyes.

As the girls spun on their heels away from me, I shouted, "That's the only way you'll ever get money, is if it comes from a man!"

The big one stopped on a dime to face me. "What?" she sneered. "You miserable shrimp-dick mother-fucker."

This made the little blond giggle. She wanted in on this too and came over to lend a voice of derision. What they both forgot is that I still had one good leg, and you should never tease a wounded tiger. I was crouching on my crutches as if I absolutely depended upon them, knowing that I had only one shot. And then I leapt.

In midair, my left hand just barely caught the little blond by the right wrist, and I yanked her toward me, throwing her high up over my head all in one smooth motion. She couldn't have weighed more than ninety-five pounds and I snatched her up like she was a child. I had my back to the overpass railing, which was streaming with rush-hour traffic below us. When I threw her, I tossed her right up and over myself, so that she flew over the railing behind me and off the bridge.

This is the part that leaves me shaking my head to this day, muttering *holy shit, what was I thinking?!* If anything had gone wrong, I would have been a murderer. Instead, in the intensity of the moment, I could feel the flow and synchronicity of life's energy all around. It was as if I knew exactly what was going to happen, as if I had my finger on the pulse of destiny.

As the girl sailed over the guardrail head first, I spun around and caught her by the ankles, one in each hand. Lord help me, if she had kicked I might have dropped her to her death. But she was literally petrified. Not a sound or movement came from her.

"Put my money AND my ID back in my pocket NOW," I commanded the big brunette. Her accomplice gently swayed above the cars below as I realized that with my hands occupied, I was basically defenseless.

I didn't know how long I was going to be able to hold on to her friend; sweat was already running through my fingers. What would happen if the brunette called my bluff? I had noted their body language after they took my money, the way they stood side by side with elbows and hips touching, and counted on my instinct being right: that there was more to their relationship than met the eye. Apparently, I was right. I glared at the brunette and without hesitation she rushed over and jammed my belongings back into my shirt pocket.

The look in the big one's dark eyes and the ball of her fists said it all as I hauled her little friend up and set her on her feet. The blond's tousled hair looked suddenly stark white and she seemed completely devoid of color. I'd never seen anyone so pale in my life. Like a specter, she dashed off without saying a word. The big one shot me a final dirty look then took off after her partner in crime. Me? I continued on my way to the discotheque to drink and dance the night away.

About six weeks later, I ran into the big brunette again. I was at the disco with my best friend, Pug, sitting at a table resting my foot while he was buying the next round. At first I didn't recognize her; the bar was darkly lit and I was wearing sunglasses. I smiled as she approached me, like I did with most girls, then she grabbed the tinted John Lennons from my face. My smiled faded instantly as I saw who she was. She then crumpled my glasses with one hand like they were made of tissue paper and dropped them at my feet. It was an impressive show of both speed and strength. Then she turned and sauntered out the dance hall.

After the disco closed down, Pug and I decided to stop in our favorite little bar across the street from the Post for one last drink. There, at a long table in the center of the Golden Nugget Saloon, was that big brunette holding court with a bunch of her friends.

I crutched to the head of the table. "You're the bitch who broke my glasses," I said to her.

She slammed both fists down on the table, making all the beer steins wobble. "I'm going to kick your ass!" she announced, springing to her feet.

Amidst the murmured "oohs," I had a sudden lack of conviction. She was taller than me, about the same weight, and really well muscled. She could easily put Rosie the Riveter to shame. Pug, not being aware of our first encounter, nor our second, was more than a little confused.

"Right, you and me outside, *now*," I demanded, trying to regain my confidence and the upper hand.

When I reached the center of the deserted main street, I handed Pug my crutches and turned to face my opponent while balancing on one foot. I got the feeling she was at least as drunk as I was and figured my key to victory lay in antagonizing her.

"You're too stupid to be Godzilla, and too ugly to be King Kong, but you sure are one big, ugly, stupid cunt!"

The anger in her eyes pricked like daggers. She came at me with a haymaker that would have killed a cow, if it had landed. At the last split second, I ducked under her swinging fist, and in one fluid motion turned to Pug, who handed me back my crutches as if on cue. We started for the far side of the street while the butch was still in motion. She had swung so wildly that she lost her balance and did a complete seven-twenty before landing hard on her backside. I wondered if she had fractured her tailbone as she sat there motionless in the middle of the street, stunned into silence. I had knocked her down without ever laying a hand on her. Pug and I swiftly turned and strutted through the sentry gates back to the kaserne, laughing all the way to my room.

Hey Harvey,

How's it hanging? I know—a little to the left!

Guess what? My Top Sgt. says I could be promoted by the end of the year. I think he's just throwing me a bone for being "dead-lined" —what a piece of machinery is called when it's inoperable—because get this. My foot got smashed by a falling road wheel, like a sledgehammer dropped on a sponge cake. Fortunately, I don't remember much. But I do remember vividly when it happened—and I'll never forget that 2nd Lt. who found me.

A helicopter picked me up in the field and took me to the hospital—how cool is that? The doc took one look at me and told the nurses to take off all my clothes and hose me down. You know that's my kind of foreplay! I spent five days in the hospital, and at first I totally felt out of my element. But hot meals in bed, nurses (say no more), and TV with a VCR won me over. I figured they'd incinerate my clothes, but they actually washed them and returned them folded, if you can believe it.

It's funny though, when my Top came by, he said they thought I had run away, which is why no one looked for me. AWOL—in the middle of the woods? Like, where the hell was I going to go? And my top sergeant when along with it? Fuck them all.

I've had time to think about what I'm going to do after this. I don't know if I'm being discouraged or realistic, but I think maybe Grandma was right. Maybe Army life isn't for me. Still . . . if I get promoted like my Top said, I think I can make sergeant within a couple years—and then I'll stay. At my age, most of my friends have graduated college and have a career lined up. I'm pretty good at what I do here, and it's a respectable profession. Hell, maybe my next assignment will be better.

I heard from my buddy Big Mike from HQ the other day. He sent me a letter from Leavenworth—The Big House. Mike says he's with mostly officers, and it's the best he's been treated since enlisting. The guards don't yell like sergeants do, and they don't make his life hell on a daily basis. I can't imagine. He makes wooden toys for kids in the morning and has free time in the afternoon. And he said the chow was really good, way better than back here.

Turns out he was bouncing for a whorehouse in Amsterdam and when his money ran out, he just turned himself over to the US Embassy. I guess they called him a deserter and he thought they were going to shoot his sorry ass. Can you believe that shit? They pulled their guns on him! Mike just wasn't cut out for this, but buyer's remorse is a son of a bitch when you've signed up with Uncle Sam. The bottom line is: nobody likes doing what we're doing —but we do it anyway.

Did you know that drugs are legal in Amsterdam? Mike wrote me about these coffee shops where you could get your choice of seriously fine hash or weed with your cappuccino—I am going there as soon as possible. But I still haven't made it out of country.

The rest of the battery gets back next week, and the kaserne is like a ghost town. I'm stuck in a room healing up with Martinez, a burly Filipino with a badly infected tattoo. I have to stay in bed with my foot up while Martinez just lies in his cot and pouts—he hasn't said ten words this whole time. He should have taken a lesson from our culture. Can you imagine doodling on yourself permanently?

I can see how for certain peoples, like Islanders (not the hockey *team, you dumb Canuck), it's a rite of passage and part of their tradition, like a bar mitzvah is for us. But a girl with big titties, wearing nothing but an overseas cap, slowly fading on your shoulder? Now that's something completely different. It just seems a little shortsighted to me. Is he still going to want that thing on his arm in forty years? Maybe by then he'll pay someone to take it off.*

*I did see a really cool tattoo when I was at the hospital, on this guy who worked there. His arm was missing halfway up the forearm, and around the edge of the stump was a dashed line with a tiny pair of scissors and the instructions: "cut along the dotted line."*

*Well, with that lovely image I'll say "Gute nacht." That's good night 'round these parts.*

*Love, Adam.*

# NINE

## Rice and Fish

The one thing I missed about headquarters was my Sunday ritual with my old roommate, Merrick. When I was first deployed to HQ, Merrick had just been busted down to private first class from specialist for going AWOL. He had a bit of a breakdown and couldn't deal with his life in the service anymore. Though he had limited options, he had come back on his own. We just happened to be roommates, which made me, the newbie who didn't know anyone else, his only friend by default. Even still, a guy going AWOL isn't the sort of thing you ask about; you wait till the person opens up to you, and Merrick wasn't ready yet.

Both Merrick's father and uncle had served in the Army, and they had come home with captivating tales of daring exploits and lasting camaraderie. Naturally, Merrick thought Army life would be full of that same excitement, yet the only stories he had to tell were of stifling monotony and overbearing sergeants.

On Sundays, out of sheer boredom, we would walk to the park at the center of town, stopping along the way to buy a gallon of cheap red wine from the little Turkish market. Merrick would also pick up a day-old loaf of bread for the ducks. We'd sit on one of the park benches by the pond and take turns forcing each other to drink—the wine was that bad. Merrick was happiest when he was sloppy drunk and throwing bits of stale bread to the birds. The birds loved him for it. I think it made him feel needed, and it saddened me to think that was the closest thing he had to genuine affection in his life. I hated watching him try to drown that fact out every time he poured more of that vinegary wine down his throat.

Our sole purpose was to drink the day away, if for nothing else but to lose one insufferable day out of the seven. And if we took a nap on the grass like bums, so much the better. Most places were closed on Sundays, and our hangovers would hit within hours unless we put something in our bellies to soak up the alcohol. So, late one overcast afternoon we stumbled upon a little Korean restaurant. As we finished our meal the owner came over, introduced himself, and sat down. He was depressed about the quality of rice he could buy on the German economy and, after several sakes on the house, wondered if we could help him out.

The restaurateur proposed that we come the next Sunday for an early-afternoon dinner, with a couple hundred pounds of rice in tow. He promised to double our money and sweeten the deal with all the rice wine we could drink and all the bulgogi we could eat. He ended the proposal by comping our meal.

To Merrick, "black market privateering" sounded like an adventure; to me it sounded like free food and pocket money. Either way, who cared if it sounded crazy. The food was good and the price was right. So the next Sunday morning, Merrick and I showed up at the big PX of the Pioneer Kaserne, Hanau's main post, to buy a couple hundred pounds of rice.

This became our weekly ritual: buying rice in mass amounts and loading it into the trunk of a taxi. "Our buddy's having a big wedding," I would say to the clerk as I hoisted the fifty-pound bags.

The complete apathy on the cashier's face was always reassuring. No one ever looked at us twice.

Shortly after I was sent to the gun battery, Merrick was transferred stateside. I maintained the deliveries but found it difficult to do single-handed. Plus, being a Canadian Jew in Uncle Sam's Army selling California rice to Korean restaurants on the German black market was as close to an international criminal as I wanted to get. When my foot got injured, the side job simply became too arduous and all of it came to an end. I have no doubt Merrick and I were just a link in a long chain of GIs who helped give that Korean restaurant their competitive edge.

While the rest of the gun battery was still in the field, I had a lot of time on my hands. Luckily, I didn't have to stay in my cot during the day on the weekends. There was a small gravelly shoal in a wide curve of the Main River as it skirted Hanau, about a quarter mile off the road, just down from a rednecked Bavarian beer garden Merrick had taken me to once. I bought a little folding chair, a fishing pole, and a tackle box, and this little jaunt became my new Sunday morning ritual. I'd load up my gear in my backpack along with a six of beer, a small baguette, some dried salami, and Almond Rocas for dessert, then catch a cab to the German cowboy bar on the outskirts of town. The cabbie always thought it odd to drop me off there. I can't blame him, considering the place would still be closed, and I'd be on crutches with my foot heavily bandaged.

As I'd slowly make my way down to the river, I would be reminded of my childhood and begin to rehumanize. I don't know when a person stops feeling delight in turning over a stone by the water's edge just to see what's underneath, but I hope it doesn't ever happen to me. Great big barges would come floating down the languid Main without fret, and I echoed their sentiment in full recline with both feet, and the beer, chilling in the cold water.

Heaven can be a line cast out with nothing to do but ponder.

I didn't catch a single fish, but that wasn't the purpose, and I was never disappointed. Afterward, once the sun had set, with my beers drunk and the last of the Almond Rocas savored, I'd amble up to the Deutsch honky-tonk on my crutches and continue drinking until I could no longer stand to listen to Hank Williams in German. Then, feeling blessed despite my delicate condition, I'd catch a taxi back to the post.

Months later, after seeing me head out with my line and pole the previous weekend, several of my friends asked if they could join me. It was a nice summer morning, a good day to be curious, and I gladly brought them along. We picked up our supplies and headed for the river where I quickly discovered it's easier to be quiet and still when there's only one of you. By the end of the afternoon, we had done a lot more drinking than fishing. The heat and humidity enveloped us, as if the earth's rotation had ceased, leaving us steadfastly stuck in the sweaty doldrums between time and space. That's when I noticed, through the warm river haze, the activity across the way.

Four cute German girls—sirens, if I had known the truth—were setting up two little tents on the other side. They waved hello to us and we waved back. It seemed to me like a gift. All we had to do was make sure a barge wasn't on its way—they might have been slow but they were so damn big—then swim across the river and introduce ourselves. One of the other three guys was not up for the swim across the wide oxbow and tried to talk us out of it. But I was young, dumb, and full of testosterone, and that little stretch of water was not going to keep me from the promised land on the other side.

Stripping down, I carefully folded my clothes into a pile. The other two looked at me doubtfully but followed suit. We were too far away for the girls to see detail, but they could clearly see we were buck naked. The chill of the water stopped me thigh high, but the girls waved us on and that was all the encouragement I needed. Doing a side stroke with my right arm, I held my clothes up and away from the water with my left hand. As we neared the halfway mark, both of my companions stopped.

"I can't make it," one grumbled.

"Neither can I," gasped the other.

"But we're already halfway," I argued. "It's the same distance no matter which way you go."

They gave my reasoning a full two seconds before turning around and swimming back, leaving me to continue on my own.

The guys reached their riverbank about the same time I reached mine, only they scrambled up to the bar while I strolled onto the shore to make some new friends. The girls were very welcoming. It's always a good sign when you approach a woman in your birthday suit and she hands you a beer. My foot, though it was healing well, still looked a little abnormal, but fortunately it didn't stand out as much as another part of my anatomy that had captured the young ladies' attention. I needed no arm-twisting to spend the night. Perhaps it goes without saying that I got to know all four quite well.

Less than a week after, a report in the local news announced a drowning on the Main not far from where we had been. Witnesses on the river's edge said it looked like something had pulled the swimmer under, and the news revealed that over a dozen people a year died swimming from one side of that river to the other. That summer alone three GIs drowned, and authorities speculated that an undertow was to blame. Armed with that information, I wondered about the girls' intentions and never attempted that swim again.

But something about the story nagged at me. I didn't feel any undertow as I swam across, and it would be highly unusual to encounter one in the middle of a long, meandering river. What is not unusual in those waters, however, are the wels catfish. These apex predators can not only grow over ten feet long and weigh over three hundred pounds, but they are über aggressive and have historically been found with the remains of large animals inside their cavernous bellies.

My guess is that those guys went missing because big fish eat little fish.

And for reasons I'll never know, the wels let us little fishies go that day.

Hi Harvey,

Haven't heard from you in a while . . . did you break your hand
in karate? Write me—you lazy bastard! I've been thinking more
about going to school after this, and physical therapy might be a
good way to go. I've been busted up enough myself and that helps a
person gain a good level of empathy. That, and the more injuries
you have, the more you learn about the human body—so I must be
more than halfway through anatomy 101 by now.

You know, not enough doctors really understand what an
injured person's going through. All the MDs know is what they read
in books, and you can't explain the depths of pain through little ink
squiggles on a piece of paper. To know what a patient is going
through, to be able to truly empathize, a doctor would need similar
life experiences. But they don't. So they have no idea what my 8 is
on their pain scale of 1–10.

What do they call the medical student who has the lowest
passing grade? Doctor! And who do you think the Army attracts, the
best and brightest? Oh no—desperadoes, like the rest of us who
couldn't fit in anywhere else. Freakin Army doctors are the worst.

But I can't complain about everything, I ran into one of my
nurses at a disco in Frankfurt the other week. Ingrid was one of
three nurses who took care of me. She's a really sweet girl, strong
but gentle, and she's teaching me German. We're starting with body
parts. I like to work from the bottom up and naturally, I'm a
cunning linguist—if you know what I mean (wink, wink, nudge,
nudge). It amazes me how aches and pains can simply disappear
during sex. Never underestimate the healing power of a good
woman's touch . . . and never hesitate to run like hell from the bad.

*Do you know the singer Sade? That's what Ingrid looks like: dark eyes, black hair pulled back, full red lips, fuller hips. So rare to find a woman both beautiful and sexy. I think her folks are Moroccan. She speaks French, but she grew up here in Germany. Her father insisted she have a very Deutsch name so she wouldn't be treated any different—at least not on paper.*

*We were talking about how the French call an orgasm "le petite morte"—the small death. That's a really weird way of looking at it. I don't think I've ever been closer to another person, or to God, than during an orgasm. It's also when I am most in the moment . . . appreciating the present without concern for the past or future. And it helps renew one's faith that life is worth living.*

*On a purely holistic level, I think everybody should orgasm at least once a day as a reminder to our bodies that we are still capable of carrying out our prime biological directive—reproduction. And as such, it's an indicator to the immune system that we are worth the trouble of preserving. You know what the Grateful Dead say about loving the one you're with? Well, that means even if you're all by yourself!*

*Okay, enough about sowing the seed—'cause if I start to think about the fields I'd like to be plowing, I won't be able to roll over in this damn saggy little cot. I'm telling you, who the hell put this damn kickstand in the middle of me anyway?*

*So study hard, otherwise you could wind up sleeping outside in the cold like me. See you in the funny pages.*

*Love, Adam.*

# TEN

## Pug Defender

Leone D. Puglia was with the communications detachment and bunked with the rest of his squad's enlisted in a room just down the hall from mine. In Italian the name Puglia is softly poetic, though not in the least in a phonetic Army parlance. As most sergeants mangled his last name, for the sake of dignity if not brevity, Leo simply became known as Pug.

The day Pug and I were introduced was shortly after being transferred to the gun battery. We had both been pursuing the US soldier's most common pastime, drinking alcohol. Immediately upon meeting, something odd happened: we gave each other the Three Stooges "secret handshake." Neither of us had ever done that before, and we certainly hadn't planned it. It was in that moment that we both knew we had a connection far deeper than our current incarnations were aware of—and a slapstick one at that—showing us that our friendship was undeniable.

Similar to the situation with Merrick and me, the ranking specialist in Pug's room thought of Pug as his best friend—but it wasn't mutual. The specialist wasn't a bad guy, just leaned heavily toward the boring side . . . the kind who found watching bowling on TV exciting. He certainly wasn't the hell-raiser that I was. Regardless, though, I felt like I had stolen someone's best friend, which can be an uncomfortable situation, especially if you're constantly running into the offended party.

From the moment Pug and I met, we clicked in a way neither of us had experienced before. Most times it felt like we knew what the other was thinking, and we could finish each other's sentences. He was the Harvey Korman to my Tim Conway; there was nothing I liked better than cracking him up. Leo was a White, Catholic, Italian-American from Philly—not the dark, swarthy kind I resembled, but more like James Caan's Sonny from *The Godfather*. Not once did Pug ever look at me any different for being a Jew, and after a while, I barely noticed his lack of skin color.

I still don't quite remember, yet will never forget, the time Leo and I were found early one Sunday morning passed out side by side at the top of the fourth-floor stairway landing. When they tried to rouse us, Pug suddenly sat up and warned, "I am the Gate-Keeper." To which I bolted upright and chimed in, "I am the Key-Master." The fourth floor was vacant and apparently we were making sure it stayed that way.

And then there was the strange psychic connection we had while playing cards.

To pass the time while on alert, I had fashioned a homemade hacky sack from scrap pieces of leather and filled it with little bits of plastic. It was an unusual activity, and not in the rule books, but the commanding sergeant let Pug and me kick it around as long as we stayed right outside the one and only door. It always seemed to be cold and damp in Germany, and one time, while going for a stray shot, Pug slipped at the door's threshold and went down hard, giving his head a good knock on the way. The knot on Pug's head was about the same size as our hacky sack, so we prudently called it a day and went back inside to play hearts.

Just as the first round was being dealt, Pug said, "There it is."

"What is?" I asked.

"The queen of hearts," he replied matter-of-factly.

The dealer grabbed the facedown card and flipped it over—and it was the queen of hearts!

"How did you do that?" I asked.

Pug shrugged. "I don't know."

"Re-deal," shouted the owner of the deck.

The next round, and all the rounds after that, Pug could tell where and when the queen was dealt. In fact, for as long as I knew him, Pug was always able tell when the queen of hearts came out in any card game. No other cards. Just that one.

That fall must have shaken something loose.

A year into our friendship I met Leo's father, a retired Army sergeant, when he came to Germany for a visit. He was a down-to-earth kind of guy, tough as nails and completely genuine. He took us both out to lunch, and as we sat there talking I realized that, as with his father, the Army was Pug's career. I was strictly a two-year man, just in it for the educational benefits afterward, but Pug was a lifer. It suddenly dawned on me that my influence was detrimental to Pug. Sure, we had a lot of fun together, but I was not helping him climb that ladder to success in the service. The truth is, the guys who were in a position to promote Pug were also influenced by who he hung out with most—and that person was me.

———

After my incident with the road wheel, I had almost a month to kill before the Death Dealers returned to the rear. Recuperating with a crushed foot is bad enough, but being stuck in the Army at the same time only made the situation worse. I had no medication and no physical therapy, only orders to lie in bed all day. Studying the second hand on my watch I could clearly see time slow down with every tick. It was as if every moment stretched out and lingered, to the point that I could feel the rotation of the earth cease and time stand still. It didn't seem rational, and yet . . .

When the rest of the unit finally returned from the field, I had something to break the tedium, and hooked back up with Pug. My right foot was still grossly swollen and wrapped in tensor bandages, and I was on crutches, but I wanted to show Pug the bars I'd been going to and the women I'd been meeting—one for him and two for me.

The first night back, Pug and I set off to the club to get properly drunk before trying to dance, like any sensible young man does. But Pug hadn't had a drink for over a month, and Germany makes some strong brew. After a couple of dopplebocks, the fatigue started to catch up to Pug and he put his head down on the table as we were talking to two lovely fräuleins. I nudged him under the table and Pug started to snore in response. The girls took it personally and, taking umbrage, got up to leave. I tried to follow them but was hampered by my crutches in the packed bar. So I gave up and returned to our table where I found a GI pestering the passed-out Pug. This jerk was slapping Pug's face and telling him to wake up, but Pug didn't move. As I closed in on the table, the guy looked up, saw me coming, and took off, blending into the crowd.

It didn't make any sense to me why a guy would bully someone who is incapable of defending himself. Even when sober, Pug was not an aggressive person. He didn't get into a single fight the entire time I knew him. Whatever the case, the antagonist had left and I spotted the girls we had been talking to. Pug looked pretty comfortable where he was, so I hobbled over to the young women, hoping to win them over with a little sympathy.

No sooner had I left than the son of a bitch who messed with Pug came back over to his table. He looked around quickly, then kicked Pug's leg. Pug was fast asleep and gave no response. From across the room I watched this guy pick up Pug's beer stein and start drinking. I couldn't believe what I was seeing and didn't even bother saying goodbye to the girls as I stormed over to the table, cursing my fucked-up foot with every awkward hop on my crutches. But I wasn't fast enough; the asshole saw me coming again and disappeared into the crowd—this time with Pug's beer.

It was already late, and I had pissed off the girls so there was no going back to them. I figured it was time to get Pug back to the post where he could properly catch up on his sleep. So I roused him awake enough to drag him outside and wait for a cab. Within minutes, the jerk reappeared. Apparently, he was not intimidated by zombie Pug's crippled bodyguard and gave Pug a shove from behind, pushing him almost off the curb into the street.

"What the fuck is your problem, asshole?" I shouted at him.

He just laughed. Then he punched Pug in the shoulder, all the while staring at me like some psycho.

That was it. He had flipped my switch into overdrive and I went straight to insane caveman. Throwing aside my crutches I jumped, taking him down like a lion does to a gazelle. I was so furious that I was gargling profanity as I pinned him underneath me and began punching his face into the back of his head.

Suddenly, I was heaved up and off him by four Polizei who put me firmly in their grasp. Knowing they didn't like American servicemen, I decided in my limited state not to put up a struggle. That's when I noticed they were staring at my foot. The bandages had come undone and it looked hideous.

After an uneasy glance between them, the two Polizei holding me let go and the other two grabbed the bloodied guy I was grounding and pounding. They yanked him off to their police car, while the remaining duo picked up my discarded crutches. Their kindness more than surprised me; I was amazed that I wasn't being arrested and wondered how much had they seen.

It wasn't until Pug and I were sitting in the cab that I noticed my pants were blood-soaked at both knees. A takedown with a busted foot on a cement sidewalk is never a good idea, and the landing was a lot harder than I anticipated. But there had been no thought process involved. I simply did what I knew how to do best.

Nobody was going to mess with my buddy, Pug—not on my watch.

*Dear Harvey,*

*What's it like having plumbing and electricity? Hell, what's it like having an indoors? 'Cause when you don't even have four walls and a roof, everything else is up for grabs too.*

*Speaking of up for grabs, did the folks sell the house? Do you know what happened to all the stuff in my room? My clothes? If they hadn't sent me the new address I'd think they were trying to give me the slip. Moving on up to the North Side, sounds like the theme of a TV show. What's the new house like? Did they buy a smaller one so we couldn't all move back in? It's weird how things happen in the real world and I feel like I'm watching it on delay from inside a bubble. I'm starting to wonder what happens when the bubble bursts.*

*I'm doing all right. I'm limping around but it's close enough to be called walking, and I can get my boot on, though just barely. The doc says my foot will get better faster if I stay on it—but by the end of the day, I just want to put my feet up. As we say in the Army, "My dogs are barking."*

*Life in the field becomes so primitive, and when the sun goes down so do we—unless we're close enough to a small town or village. The officers almost always stay in one of the local inns, so they're not around. Without officers to fuck things up, sergeants tent with themselves and just leave us enlisted alone.*

*When we're bivouacked in the German countryside, all these little kids from the closest village will ride up to us on their bikes just before twilight. These Wunderkind, big and little, come loaded down with backpacks full of all sorts of odd bottles of alcohol. I'm guessing the hooch is whatever they could steal from their parents'* cellar: dusty old schnapps, weird herbal liqueurs, and musty bottles of Monk's beer in exchange for a few marks and maybe a trinket like a 50 cal casing. Then we sit around a flashlight and drink with shadows crowding in from all sides. It's amazing how small the world gets without proper lighting.

Funny that the villagers are so accustomed to seeing our machines of war that their children profit off us. I bet this goes back to the First World War. When a country gets beaten and goes broke, the only currency left is booze—but that's just because you can't put pussy in a bottle!

I'm pretty sure all of man's major conquests have been about getting a piece of putang. Look at Ghengis Khan: he sired an entire generation. And Columbus? What was he looking for, and what was he hoping he'd get when he came back home? The same thing we all want! And Ponce De Leon just wanted to find the fountain to be young enough to get it up again. Fortunately, of all my problems, that's not one of them.

By the way, sorry for this crappy little notebook paper. It's all I have right now. Like most of the guys, I keep a pen with me and carry a folded wad of paper in my pocket. Whenever I get a chance and the feeling is right, I write to you. So these letters are usually cobbled together from a couple of days, or even a couple of weeks' worth of 10 minutes here and 15 minutes there—often written in the very last light of the day, or while waiting in line.

And as long as I'm apologizing, I won't be able to send you a check this month. I bounced one and just learned that I have 17 (yeah 17!) more outstanding—which is usually a good thing. But this time it's not. Motherfucker! I didn't realize how much Uncle Sam was taking out. When I told that to my Top Sgt., he just laughed.

*Then that old son of a bitch gave me the Article 15. That's strike one. Two more and I can get: 1) thrown in jail 2) kicked out 3) all of the above. Well, there goes my perfect record . . .*

*It's really not that big of a deal, though. This is the most common offense for enlisted like me who aren't used to having a checkbook. So . . . it's not unusual, but I'm broke, which bums me out. No mun', no fun, and I gotta run.*

*Wishing I was there (ain't you glad you're not here?),*

*Love, Adam.*

# ELEVEN

≈

## Head Banger

The unpredictability of life in the military is like playing hide-and-seek with monsters: you never know when something terrifying is going to suddenly jump out at you from the darkness. Sometimes the malevolence slides right by, barely brushing up against you, as it goes after its next victim. You cower in its shadow, hoping it doesn't notice you, and think, *There, but for the grace of God, go I.* Other times it's like the monster has your scent and is actively seeking you out. You can try to outrun it, but the horror is always waiting for you, when and where you least expect. And sometimes, worst of all, you have the stark realization that the monster is you.

A bunch of us were going out together one Friday night, and as we passed through the Unit Police shack to the street, the two men on duty detained me. I was off the crutches but still limping badly, so I was lagging at the rear of the group. My friends up ahead of me paused.

"Go ahead and get a cab," I said, waving them on. "I'll be right there." I didn't know why these U.P.s were stopping me, but I was sure it was just routine.

As the last of my friends walked out of sight toward the taxi stand, I saw the U.P. by the front door turn and lock it. Unfortunately my spidey-sense wasn't tingling that day, and my focus on this guy meant I wasn't paying attention to the second U.P. at the rear door. Before I knew it, he had come up from behind, grabbed me by the shoulder, and spun me around. I don't know what I was expecting, but it wasn't getting clocked upside the head.

With no warning or reason why, this guy clubbed me with his Mag-Lite just above the left eye, splitting my brow wide open and spattering the walls of the shack with crimson dots. I didn't even know there was a fight in store for me—and all of a sudden I'm not just in the middle of it . . . I'm "It." I felt blood cover the left side of my face as the brute by the front door took a swing at me with his baton like I was a freaking piñata. In a flash, the two of them were double-teaming me, and I quickly realized what a trap I was caught in. It was then that survival mode kicked in. I knew I had to make it out of there before I got knocked to the floor. If I went down, I was sure I wouldn't be getting back up.

I bull-rushed the guy behind me, slamming him into and through the back door. We spilled out into the entryway and I jumped to my feet. His buddy was there in an instant. I threw a couple punches to back them off, but I could barely see for all the blood in my eyes. I was clearly on the losing end of this skirmish, and the reptilian part of my brain urged "retreat," so I started backing away toward my barracks. The U.P.s just stood there, holding their ground by the gate, with shit-eating grins on their faces like they had done something to be proud of.

If they thought handing me a beat-down was that easy, they had another thing coming.

Masked in blood from my head to my waist, I startled the guy at the front desk of the barracks.

"What happened to you?" he asked, wide-eyed.

"Got beat up by the U.P.s at the gate," I replied blankly.

He was rendered speechless as I continued toward my room. Once inside, I headed straight for the bathroom. I was making good use of my first-aid kit when I started to plot my revenge. I was guessing my friends had gone ahead without me and had no idea I was attacked. The whole thing happened so fast that I had probably staggered away by the time they double-backed for me. By then, who knows what the U.P.s might have told them.

More than anything, I felt embarrassed for letting those dumb-asses work me over like they did and ashamed of myself for such weakness. Injury to insult, the corner of my head where I got hit looked like a tomato that had fallen on the floor. I was in need of some serious mending, but if I wanted to settle the score with the U.P.s, I'd have to do the mending by myself.

Now, I don't recommend anyone try to sew themselves up, but if you should be required to make such repairs at home, then make up your mind fast. Shock is a gift from nature but has a short shelf life—use it or lose it.

One of the first things a recruit is issued is a sewing kit. The cut above my eye wasn't more than an inch long, but it was deep; luckily, the massive gush of blood that had poured from it had no doubt washed it clean. I sterilized the needle with a lighter and soaked my thread in Bactine. After threading the needle, I ran it and the thread through a gob of Neosporin held between two fingers. I knew not to try to sew a seam like I was mending fabric, but instead to go at it more like tack welding. That method would help the wound stay closed, and it would seal by itself. I gritted my teeth through the three sutures. The first one in the middle of the gash was the worst, then those on either side of it went a bit easier. I used the same practiced stitch I had taught myself while sewing up my homemade hacky sack, finishing each suture with a solid double knot. It not only came out nice and neat, but the black thread practically disappeared into my dark eyebrow, leaving little trace.

Both my shirt and pants were soaked in blood, and I didn't want any mementos from that evening, so I shoved them into the garbage. I didn't know why those guys had attacked me, but I knew I couldn't let anyone think they could jump me like that and laugh about it later. The Army is a dangerous, testosterone-fueled playground. Let yourself get punked once, and chances are it'll happen again.

I was due to go on duty guarding the off-base military housing at twelve that night, which meant I had to go back through the guard shack. Normally Slim would be joining me, since we pulled that shift together, but I gave him the night off. I knew he'd stick up for me if he was there, and this was my fight. I didn't want to risk him becoming collateral damage. So, cleaned up, stitched up, and completely sober, I made for the guard shack in my battle dress uniform, ready for war.

My assumption is they never saw me coming. I kicked in the back door and took only two long steps before I was on the first U.P. I grabbed him by his flak jacket and threw him laterally through the side window. He fell with a glassy smash onto the driveway outside and lay there motionless. Then I made for the son of a bitch who split my skull with his flashlight. I saw the terror in his eyes now that he no longer had the element of surprise on his side or his partner for backup. Instead of taking me on, that yellow-belly took off running like lightning.

The Good Lord saved me that day. If not for my limp, I would have caught up to that coward and surely killed him with my bare hands. My anger at that point was so primal and absolute that I was devoid of all empathy. I had become a monster.

My heart screamed and pleaded with me to catch him and

drub the life energy right out of him. But he made it up the steps of the U.P. building just ahead of me: the Sergeant Major's House and home to the rest of the police force. I stopped at the bottom step as the U.P. stopped at the top, halfway through the front door. We both knew I couldn't go in there—it would be suicide. But I had no plans to leave until he came back down to face me.

Eventually, someone called my top sergeant and I was remanded to my billets until the next morning, when the sergeant major would be present. He ran the entire post, and the military police were under his wing. An attack on them was an attack on military justice and would not, could not, be tolerated. Thankfully, my top sergeant was a decent person and a tough old bird who wasn't easily pushed around.

That next morning, Top and I met with the sergeant major. The two chicken-shit U.P.s were absent, and it was decided that because the U.P.s had started the fight without provocation and offered no explanation for it, my retaliation was seen as tit for tat. One broken window in exchange for one busted head was considered even-steven, no harm and no foul. That was good old Army justice plain and simple.

I've always wondered why those two guys targeted me. Was it a random act of violence? Were they simply trying to cull the herd? Or was it payback someone had put them up to? I was certainly no angel, and I wondered if it was revenge for something I had done. But nothing specific came to mind, so I chalked it up to a case of mistaken identity. What was really odd, though, is that I didn't recognize those guys from before, and after that night, I never saw either one of them again—and believe me, I looked for them until my time was up and I left the country, pondering if the sergeant major somehow had the last laugh.

Soon enough, I would find out.

*Hey Harvey,*

*Guess where I am? That's right—my home away from home in the sticks. You ever wonder why so many veterans end up homeless? I'll bet it's partly because we're used to living outside. On top of going on maneuvers, once a month we have a simulated war situation, or what we call "When the balloon goes up." I'm pretty sure the sergeants get a forward notice. Most of them live off post, and they come busting into the billets in full gear yelling and kicking cots. I stay packed and ready; it's easier that way. And honestly, we never know if it's going to be the real deal or not.*

*The kaserne gets left deserted—everybody out and the front gates locked. In a long line of trucks and track vehicles we drive straight out of the city deep into Butt-Fuck Bavaria. Within hours we are buried in the Black Forest—locked and loaded in full NBC (nuclear-biological-chemical) gear. Every person and every bit of equipment, big or small, comes camouflaged, and it's amazing how, when we pull off the road and into the forest, this mechanized circus just melts into the scenery.*

*The piss d' resistance are these huge camouflaged nets that you could hide a house under. We set them up on poles overhead to hide our tents and drape them over the howitzers to break up their profile. We've got this shit down cold (it's not rocket science) and it takes only minutes to make the entire artillery battalion vanish. We're invisible from the air, and even on the ground you'd never know we were there until you tripped over us.*

*These forests where we play soldier are blood-soaked in history. Wars have been fought here since motherfuckers dropped out of the trees and started clubbing each other. These are the same Saxony woods that the Romans battled the Germanic tribes in, where Charlemagne and Napoleon sought conquest, and where decisive battles in both World Wars took place. I keep wishing I had one of those metal detectors with me. There's at least one solid gold Roman eagle-standard unaccounted for somewhere around here—now that would be a proper souvenir to bring home!*

*So anyway, we had to "bug-out" at 2:15 yesterday morning. And dammit . . . it's always in the middle of the night. And usually on Sunday/Monday 'cause nobody's sober on Saturday. Shit . . . why can't we ever do this after the sun comes up? It would be a lot safer. I was riding shotgun in the five-ton, because I know how to drive one, and in the early twilight, the driver pulled over so we could all take a piss.*

*The ditch at the side of the road was a little deeper than it looked in the dark, and as we pulled back onto the road, the truck began to tilt sideways. All the guys in the back got thrown to one side and I just about slid into the driver's lap. The more we pulled back onto the road, the more the truck leaned. And it's a big truck! It felt like we were past a 45° angle, and I swear if two guys had sneezed at the same time, that big bitch would have rolled over on us.*

*So everyone got out and we let the driver do it by himself. All I could hear him saying over and over was, "Oh, shit . . . oh, shit." He had a real bad angle and it was just a matter of inches before that fucking thing tumbled over. But what the hell else was he supposed to do? He had to get back on the road.*

*The sun was just coming up and I stubbed my toe on something big. I looked down and at my feet were some leaves growing up out of the dirt. I guess the farm-boy took over. I gave 'em a yank and*

*out came a 20-pound sweet potato! Do you know how big that is?
It's freakishly humongous! And heavy. I shoved it under the front
tire on the low side, got the hell out of the way, and yelled at the
driver to give it gas—and it worked! Then we all got back in and
went on our way like nothing happened. Have mercy, Miss Percy
and good golly, Miss Molly—we were saved by a veggie who gave its
life so we could roll on.*

*Well, it's getting late and I think I hear Hansel and Gretel
riding up on their bikes. I'm looking for some rot-gut German
whiskey tonight and Five-O the driver owes me a drink.*

*Take care,*

*Love, Adam.*

# TWELVE

---

## Mosquito Wings

John "Slim" Pickens and I were sitting in a smoky little brothel, Club 69, on the outskirts of Hanau. It was late November and the chill had come before the snow. It had been much milder just a few hours earlier, and we were dressed for our celebration, not the cold weather. The wind whistled and shrieked outside, angrily tapping out a random backbeat on the front door as it tried to force its way in. Slim and I had survived the insurmountable by breaking the rules, and even more importantly, escaped detection along the way. The end was finally in sight—without our respective sergeants being any wiser.

Every day for the last eight weeks, Slim and I had been assigned sentry duty—four hours on and two hours off, around the clock day in, day out. Now, sixteen hours on with eight off is simple enough, except our downtime was broken into only two-hour increments. Normally, because a person could never get more than a couple hours' sleep in a day—and even then it was segmented into little REM-less bits—this detail was limited to four days. But normal wasn't what Slim and I were assigned.

Not a lot is known about sleep, which is weird considering how big a chunk it is of our existence. But it's a fact that a person will die if they don't get enough. One's life energy, or "chi," simply starts to flatline without proper shut-eye. The guys sentenced to that hard duty looked like animatronic scarecrows after only a few days. Sleep deprivation makes every thought and gesture so lifelessly mechanical.

Our platoon sergeants each had their own reasons for trying to get rid of Slim and me, hence our being put on that detail for two months straight, which was as unrealistic as it is inhumane. And if we had been the fools they thought we were, it would have been the end of us—mentally, physically, or both.

My platoon sergeant, DumbOx, who said I was AWOL during my lost week in the forest, got chewed out for that blunder, and he was still pissed off that I had been rescued. Worse, the only other person responsible for my well-being was my top sergeant, and he was trying hard to wash his hands of me. All either sergeant cared about was that I was out of their hair.

Political tensions had been running high in the wake of a recent truck bombing in Aukar, Lebanon, one that had been detonated just ten yards outside the US Embassy annex. Of the seventy people injured, twenty were Americans, including the US and British ambassadors, and two American soldiers were among the twenty-four people killed. Just as shocking was the kidnapping of CIA station chief William Buckley earlier in March. That was after the bombings of the Marine barracks in Beirut and US Embassy in Kuwait the previous year. So security around the off-post military housing, where most of the officers lived, had been stepped up. But, in typical Army fashion, the detail ended up being used as a disciplinary tool.

After the first week on patrol, dead tired without relief in sight, I made a bold executive decision: every other day Slim and I would give each other alternating days off. Because we were no longer part of our platoon's regular rotation, our sergeants had no reason to keep track of us. So half the time we patrolled our beat through the neighborhoods and apartment buildings together, and half the time there was only one of us. The respite wasn't a lot, but it allowed us to individually rest up every fourth day—which was just barely enough for us not to be a couple of zoned-out zombies. After all, we still had a sense of duty, and it was our necks on the line as we walked the block. The real kicker, though, was that the M-16s we carried were just for show. They refused to give us any ammunition.

The one and only time we met the detail's commanding officer, a second lieutenant, was our first day on the job. He outlined the official strategy if we came across any terrorists: chances were that one of us would get shot or stabbed, and that would likely create enough of a diversion to allow the other guy to escape. We were also advised that in the event of such an ambush, we should throw our weapons at the assailants before running, with the hope that gaining our rifles would be seen as more valuable than taking our lives. To top it all off, there wasn't anyone checking up on us in case something *did* happen. With zero backup, or even walkie-talkies, it's no wonder we didn't have any confidence in the endorsed plan of action.

In spite of our accommodations, that shift was still a killer, and living without sleep is harder for some than others. My go-to remedy was transcendental meditation. I had read about astral travel when I was thirteen and had been trying to perfect it ever since. I found that twenty minutes of T.M. was equal to two hours of sleep. Slim didn't have a method like I did, though, and he had a harder time; after only the second week, he was at his breaking point. But we still had six weeks to go, so we had to be smart about giving ourselves time off without being busted.

A typical Doughboy, there wasn't anything "slim" about John, and he used a lot more alcohol than most to cope. I regularly found him half dressed and passed out in his cot just before going on patrol. I'd have to finish dressing him, then prop him up with his arm around my shoulders and walk him through the front gates, supported like a rugby hooker. The sentries never paid us much mind; we were just two insignificant schmucks on a shit detail and we weren't their responsibility. Still, we had to make it look good and, crazy as it sounds, for two whole months we did.

Slim and I had been deemed expendable and disposable, and there was no doubt that detail was meant to break and bury us. But inexplicably, throughout those long weeks, we remained un-bowed and unbroken. It was on that chilly November night of our final shift, a little past eleven and time for us to get back to the post, when three American tourists came in and sat down next to Slim.

The "club" was nearly empty. Slim and I had already enjoyed some female company and were just finishing our last beers. To the side of the bar, on a low table, was a woman of a certain age wear-ing nothing but metallic-gold sunglasses and an itsy-bitsy gold lamé bikini. A real Skinny Minnie with great muscle definition, she swayed and flexed to the Motown playing in the background. It was different than the usual go-go girl gyrations—it was more like she was going through a groovy bodybuilding routine.

While the show wasn't overtly erotic, it did have a slow, smoldering sexuality—and I was intrigued. I had rather good muscular control and, facing her, started to mirror her pulsating pecs with my own. There was no speech between us, and I couldn't see her eyes, but I felt we were communicating nonetheless. When she did a belly-roll down and then back up, I matched it, and we seemed to make a definite connection.

Through the music and smoke, I could hear the tourists talking up Slim and discerned the word "hashish" several times. John was shaking his head; if he had been holding, we would have already gotten high. I glanced at the clock and saw that we had fifteen minutes to get going, which meant that we would have to double-time it back to the post in the bitter cold. Then, inspiration hit and I thought, *Screw these senators' sons . . . touring Europe on Daddy's dollar.* It bothered me how they were trying to use us to score drugs. Hell, I was no privileged one.

I turned to the tourists. "Did you guys rent a car?"

They all nodded.

"Then how about giving us a ride back to our post and we'll sell you some hash," I suggested.

It seemed like a good idea at the time.

I didn't have any and I knew Slim didn't either, but we both needed to get back to the kaserne pronto. Those fools agreed and we hopped into their car.

Once at the billets, we jumped out and thanked them for the ride, with no intention to hold up our end of the bargain. "Later, suckers," I said under my breath, totally unaware of the karma that was about to bite us in the ass.

The guys weren't tourists. They were the CID, the Army's undercover police—and we were the suckers.

If Slim and I were paying attention, we would have noticed them parking the car instead of leaving empty-handed. But we didn't. Those sneaky agents followed Slim up to his third-floor room and ransacked it until they found a single ten-dollar gram of blond Lebanese wadded up in foil and stuck into the underside edge of his wall locker's lower shelf. I think Slim had forgotten all about it, but of course the agents found it—and then they came for me.

The door to my second-floor room was ajar so I could see just enough to get dressed without having to turn on the lights and disturb my roommates. But the CID bastards didn't care who they woke up. Kicking the door wide open, two undercover and two uniformed military police burst into my room. With revolvers drawn, they handcuffed me in front of my dazed roomies. The M.P.s led me away while the other two stayed to rummage through my locker, hoping in vain to find more drugs.

By the book, the Uniform Code of Military Justice that is, Slim and I were each taken in different cars to their "secret head-quarters" and interrogated in separate rooms for over six hours. They acted as if they were breaking up the German Connection over one gram of hash. Cynic that I am, I told them I had another fourteen kilos of Black Afghani cemented into the cinderblock wall of my room and they should go smash it open to find it. They were not amused.

Finally, before being released, we were told we had to sign documents detailing our offenses "merely as a formality." What we signed, unknowingly and without any legal representation or recourse, was our admission of guilt in the "use, possession, and distribution of a controlled substance."

All of this just so happened to coincide with the transfer of the battalion's commanding colonel; the resulting power vacuum temporarily left the sergeant major in charge with absolute and unchecked authority. That was the same sergeant major who was in charge of our post's U.P.s, and it was his signature on the bottom of the paperwork we signed. That whole sting operation was all about payback, served up nice and cold by a pencil-pushing, desk-jockey son of a bitch. Not only were we fined, but we were also knocked down a pay grade. Slim went from specialist to private first class, and I went from PFC to private second class.

That jackass may have taken time, money, and rank from both of us, but we refused to allow him to steal our spirits.

— — —

Every evening for the next six weeks, Monday through Friday from seven to ten p.m., Slim and I had the Sisyphean task of sweeping the entire post. Pug and one of Slim's roommates, Washington, would stash beers along our course at predesignated spots, and occasionally, one of us would duck down for a refreshment while the other made a distracting show of sweeping with bravado. Then we'd pop back up and continue on our way.

Every night, we finished up by gaily strutting past the billets of the unit police, Slim and I in perfect step with each other. Pushing the brooms with only our left hands, we would make grand waving gestures toward the sergeant major's quarters with our right. It was then and there that I coined the motto later adopted by communications, who spray-painted it on the inside of their APC: "We get fucked . . . but we give no satisfaction!"

I now knew why Mac the Mechanic showed no fear. You can only take so much from a man, and without anything left to lose, he becomes very dangerous.

*Hey Harvey,*

*What's the good word? I wish it was bird—as in the "Freedom-Bird." Remember way back when I thought I might get promoted? Well, it looks like I went the wrong way! HardWay Harris they're starting to call me. I just got busted down for something I didn't do. And I know everybody who gets busted has the same sad song: "I didn't do it . . . I was framed." But it's all too true.*

*They fucking busted me for use, possession, and distribution of a controlled substance—hashish—even though my piss test came back negative. I didn't have any on me or in my room, and I didn't sell any to anyone. So, how is that even possible? I've now been fined, stripped of rank, and forced to do janitorial work in the evenings after dinner.*

*The whole thing was a setup by the sergeant major—a complete asshole and a sore loser. It was payback. They were after me and my buddy got caught up in it. His big mistake was having the hash stashed in his locker. From my very first day in Germany, I knew I couldn't leave that kind of shit anywhere inside my room. This place is worse than high school. Remember when they brought in the dogs at Oak Park-River Forest? There are surprise searches, inspections, and piss tests here all the time.*

*Our buildings have a brick façade and I've always wrapped my hash in foil and jammed it between two bricks right under the window's ledge. From the ground it's pretty much invisible, and even if it was found there it would be plausible deniability. I share my room with four other guys. How could they prove it was mine? Guess my buddy didn't get the memo. Or maybe he didn't have sisters he had to hide his Halloween candy from—but now we're both fucked!*

*And you know how the Army follows Napoleonic Law—guilty until proven innocent, except they stop at the guilty part around here. Our top sergeant didn't do a damned thing for us, so he's a bastard too, and legal representation was never offered.*

*I had to go to the main PX to buy my "mosquito wings," the shoulder insignia of a private second class, for my uniforms. I swear I could feel the disapproval from the cashier. Same thing at the tailor's shop. There's no need to explain when they have to remove my PFC patch and replace it with one of a lower pay grade. Oh, the shame . . .*

*Now it'll take me at least 3-4 more years just to make specialist. And if I get one more major infraction, one more Article 15, that would be my third strike. They'll send me straight to Charlie's Chicken Farm where they practice forced rehabilitation. Fuck that shit. And fuck this place—it's a dead end for me. It's funny, when I got out of Basic I thought I had found my calling. But not anymore!*

*If I didn't have a case of the ass before, I've got a Guinness World Record Book–sized one now! Woe to the mother fuckers that cross me. They can knock me down but they can't make me stay down, and if they didn't like me before then they're really going to hate me now. What comes around goes around. Looks like I'm going to leave this Army a lower rank than what I started with. That's not easy to do, but it is typical HardWay Harris!*

*Do you remember Dad's Yiddish high school chant, "hawkem in the beitzhem"—kick em in the balls? They would start off with a whisper and get louder and louder. Well, that's what I'm shouting now. Fuck 'em all and let God sort them out . . . the sons of bitches.*

*Hey, speaking of balls, my homemade hacky sack tore a hole and I was wondering if you could send me a replacement. I'm going to keep kicking this one till it dies, but its guts are already spilling all over the place. All my privileges have been revoked, but there's nothing in the uniform code of military jackasses about hacky sacks.*

*And at very least Oktoberfest was a blast, and even if I can't leave the country I can still drink the beer. The German version of Halloween, Fasching, culminates at 11 am on November 11th by drinking the thick, heady beer dredged up from the bottoms of the barrels. The drinking doesn't stop until the barrels are clean. That can take several dedicated months—and then they start all over again!*

*I got a funny compliment the other day from an old German guy in a bar. He said I drink my beer like a German. I don't know how to feel about that . . .*

*I ran into my old roommate from HQ—he was the driver for the major and told me this crazy story. He was sick the other week in the field and his replacement slid the jeep off the road in a rainstorm and drove right into a tree. The jeep was one of those older models, the kind with the flat windshield, and when the major went through it, it sheared his head clean off—a complete decapitation. When he got his Jeep back, it was still covered in gore and he had to clean it up . . . I can't even imagine.*

*Well, wouldn't you know it, Roland just walked in with a case of cold ones, and I've got to go see a man about a beer.*

*Take care and I'll see you when I see you.*

*Love, Adam.*

# THIRTEEN

## Field Promotion

If I said I saved the lives of every man up and down the gun line
of the First and Fortieth, I would be a liar, a fool, or both. In
the same light, to say Kuntz almost killed us all is equally wrong,
though it does smack more of the truth. We may be instrumental
in the egress of the soul, in or out of this plane of existence, but
at best that's all we are—mere facilitators. To infer that a person
is able to have performed a miracle is a crime of unchecked ego,
and stealing credit from the Eternal is not wise. *Not by strength, nor
by power, but by my spirit sayeth the Lord.* Likewise, don't look for logic—
the rhyme and reason behind it all is beyond our scope. Our eyes
don't open wide enough to take in more than a small corner of
the big picture.

I never liked Specialist Kuntz, mainly because I never saw his
true face. It was always hidden behind a poorly constructed per-
sona. He was transferred to my platoon just as winter of '84 was
setting in. It can be awkward being the new guy and trying to fit
into a close-knit crew. But Kuntz didn't make any attempt to bond
with us, like he already knew it wouldn't work out. He didn't hang
out with us, or go to the bars after work and drink with us. He
was the only lower enlisted among us who lived off base, and after
five o'clock, he went straight home to his wife.

Kuntz was not only a know-it-all but he'd always "done more" and "been better," like it was a competition. No matter what the situation was, he had a story that surpassed anyone else's. Regardless of how far-fetched the tale was, Kuntz could always top it, told with an air of contemptible superiority. If you had gone to the moon in a NASA spaceship, Kuntz had gone to Jupiter in a tricked-out diamond he had fabricated himself. You could see through his darkly glinting eyes that there were weird, weaselly thoughts racing around in his head. It was as if he believed his own lies so deeply it never occurred to him that we all knew he was full of shit. He would also tell wildly inappropriate sexual stories about his wife—things none of us wanted to hear. It was a blessing in disguise that Kuntz didn't try to get chummy with us since all in all, there was just something very unsettling about him. His demeanor was best described as a combination of compulsive liar crossed with psycho killer. And he was our new fuse setter.

— — —

When my time came to leave Germany and I needed to replace lost equipment, I took Kuntz's. I didn't consider it stealing, because I took it right in front of him, daring him to do something about it. He was smart enough not to. It didn't feel good being the bully, but Kuntz had it coming. He had crossed the line, and there are some offenses that are not forgotten—and are never forgiven.

The M-110 eight-inch howitzer shot rounds that looked like giant bullets, packed with over two-hundred pounds of TNT. Those rounds were set off by a warhead that was threaded into its nose cone by the fuse setter. The rounds, or "joes," came prepackaged with a huge eye-bolt screwed into the pointy end where the explosive timer would be attached. During a fire mission, our howitzers would line up side by side a hundred yards apart, and an APC with enough rounds for everybody would pull up a couple hundred feet away.

Lower enlisted cannon crewmen, like myself, would scramble through the Grafenwoehr muck to the munitions carrier, where we would turn and back up to the carrier's chest-high deck. Someone would slide a round over our shoulder from behind, and once we had a firm grasp of it, we would hump the joe back to our respective guns.

To "hump a joe" is to move as fast as possible with it on your shoulder, holding on to the big bolt in front of you, with the heavy flat bottom behind. If you were a tough guy like me, you carried two of those enormous rounds, one on each shoulder. When people see you carrying over four hundred pounds of explosives, there's a lot less chance they'll think of messing with you. It was a brute show of machismo—something we were encouraged to do, though few were able. At the time I had no idea about spinal discs, or how easily they could rupture.

An armored personnel carrier, with its rear doors wide open, was backed up to each howitzer, and that was where the fuse setter did his work. Because the rounds were so heavy, the howitzers were outfitted with a huge robotic arm that looked like something off the Transformer's Optimus Prime. It had a basket for a hand and swung out from the rear of the howitzer to reach into the back of the APC. The fuse setter, in our case Kuntz, would carefully place a primed missile into the basket, and then the mechanical arm would load the round up into the massive gun barrel.

We dropped off the joes to our fuse setter the same way we picked them up, by turning around and, holding on to the big nose-bolt, slowly lowering the projectile down until its flat bottom reached the APC's floor. Kuntz would then roll the joes on their bases to the back of the APC for safekeeping. During a fire mission, Kuntz would select a round, unscrew the eye-bolt in the nose cone, and gently replace it with a warhead. The warhead's detonator had only two settings, quick-time and delay, selected ever so carefully by the fuse setter using a specific fuse wrench. When set to quick-time, the round would explode upon its first impact, about fourteen miles down field, picking up one square kilometer of earth, turning it over, and dropping it back down. When set on delay, the round would bounce randomly up to seven miles on its first impact like a stone skipping on water. The charge would then be detonated by the impact of the second landing. The delay setting also caused an internal timing device in the warhead to be set into motion after the initial strike to its nose, just in case it landed in something soft, like thick mud. That way, no matter what or where, the round would explode—either on its second hard impact or simply when time ran out. And that's why we always set the fuse to quick-time.

The warhead was delicate enough that it had to be screwed onto the nose cone using the outer edges of one's palms, toward the pinky side where the skin is rather smooth. That relatively small explosive device, the on switch for hundreds of pounds of TNT, was a delicate piece of machinery. To compound the gravity of setting the fuse, there were at least a thousand pounds of TNT in the extra rounds stored in the back of each carrier.

The event that brought us all to the brink of being extinguished occurred in the middle of performing a live-fire maneuver.

We were chewing up big chunks of Mother Earth almost fifteen miles down range, and I was making my way to our gun with a couple of joes. As I was lowering my two rounds gently on the deck of the APC behind me, something compelled me to turn around. Kuntz had set the warhead's fuse and was lowering the armed missile into the robotic arm's hand when it appeared to just slip through his fingers. With a tremendous clang, the round fell nose first onto the carrier's metal floor. Kuntz had literally dropped the bomb.

Absolute stillness fell over every person on every gun. No two guys were off in the corner smoking and joking; there wasn't some guy picking his ass and looking around confused. It was as if everybody was instantly aware of what had happened. Nobody panicked and nobody yelled; we all knew this was not something that could be outrun. So we all just stood there silently, almost peacefully, enveloped in a tremendous pall of serenity, of surrender to the inevitable—the Eternal—waiting for our world to come to an end.

For what seemed like minutes, though it couldn't have been more than a split second, I remained motionless as if hypnotized. Then I snapped out of it.

Kuntz was still standing hunched over the fallen joe as I leaped up into the back of the carrier. I grabbed Kuntz by the scruff of his neck and his belt and heaved him off the APC into the frozen mud below. Scooping up the round, I loaded it onto the howitzer's mechanical arm and yelled at Roland the Gunner, using the voice of command to break him out of his trance. He immediately manned the controls and the round was automatically loaded up. Then, just before firing it off, he gave me a quick look.

Roland knew as well as I did that the only reason the round hadn't gone off was that Kuntz hadn't set the fuse to quick-time. He had set it on delay. By dropping the round, Kuntz had just activated the auto-timer of the warhead's fuse. The mechanism had been set into motion, and it was not a matter of if, but when, it would explode. The problem was, to send over two hundred pounds of TNT more than ten miles down field required a sizable explosive charge shoved right up against the missile's bottom.

That charge would, no doubt, be equivalent to a second impact and chances were it would detonate the round in the gun's tube. Such an explosion would set off all the extra rounds in our APC, and the resulting firestorm would spread to all the other howitzers in a cascading catastrophe that would kill everybody up and down the line. Both Roland and I knew that, but the clock was ticking and either way we were damned.

Roland pulled the lanyard to fire off the round. But the earth-shattering event that should have happened didn't.

Instead, a minor miracle occurred. The lanyard clicked, the round was fired off, and it safely detonated down range without a single casualty.

An empirical truth was revealed that day to all of us: the one person a gun crew must be able to trust above all others is the fuse setter. On a good crew, the gun sergeant is easily expendable and the gunner can be anyone who can pull a lanyard. But only the setter gets to test the boundaries of who plays God. Only the setter can hold everyone's lives in his two hands. And when the fuse setter can't be trusted—that's when he gets replaced.

It turned out that a lot of pressure got placed upon the fuse setter, which explains why Kuntz did what he did. A few months prior a similar situation had occurred: a fuse setter had been coerced by the rest of his platoon to set the warhead to delay on purpose. That round skipped on its first impact instead of immediately exploding, and then it leaped five miles in an unexpected direction before landing. Upon its second impact it detonated, killing a squad of German regulars towing a broken-down track vehicle into the target area. The issuing howitzer's gun sergeant was the one held accountable and, from what we all heard, got sent straight to Leavenworth.

Now here we were with our gun sergeant, DumbOx, who was

as widely reviled by all those below him as those above. Pressure had been put on Kuntz to miss-set the fuse in order to get rid of Sergeant DumbOx, but Kuntz had had a sudden loss of conviction and couldn't go through with it. His solution was to kill us all.

That was the moment we all knew Kuntz couldn't be trusted and our fuse setter would have to be replaced lickety-split. I wasn't a contender for one glaring reason: the fuse-setter was a job for an E-4, a specialist's spot, and I was half that rank—a lowly E-2 private. But because I had already proved I wasn't susceptible to intimidation, the entire gun crew, including Sergeant DumbOx who hated me as much as I hated him, insisted the job go to me.

From that day forth I was the Death Dealer's fuse setter. It may have been an unofficial double-jump-step field-grade promotion only in effect during fire missions, but it was the closest I would ever get to a promotion in the Army. And I didn't have to hump joes anymore. They gave that job to Kuntz.

*December 10, 1984*

Hi Harvey,

Well, another year older and closer to death! How's life kicking you around? Have you sent me a hacky sack yet? If you send it much later I won't be able to call it a birthday present. I was actually thinking a granny sack might be best out here. It's always wet and cold, and the crochet should hold together better than cheap pleather. Maybe you've already put it in the letter I know you're writing at this very moment?

I just got an unofficial promotion as the fuse setter. It's cool . . . at least I don't have to worry about some other idiot blowing us all up. My first time priming the warhead, my buddy hands me a piece of chalk. I didn't know what I was supposed to do with it. I thought it was for my hands, but he tells me to "Draw on it." Yeah, on the 200 lb missile we're going to shoot. Just when you think it couldn't get any crazier—there's one more layer.

The only thing I could think of drawing was a naked woman straddling it like she's riding a horse. So now every time we fire off a round I draw a slightly different Lady Godiva humping the joe. And I gotta say I'm getting better—hair blowing in the wind, thigh muscles taut and nipples at the ready. And I've got some colored chalk, so some days she's a redhead and some days she's a purple punker.

I'm still on suspension and can't leave the country, so the rest of Europe will have to wait. But I'm getting to know Munich like the back of my hand—it's a bar monkey's wildest dream come true (and it's really good to get out of Hanau). Munich is a city built on beer, and Spaten, brewed for the working man, rules. But my favorite is Pilsner Urquell. It reminds me of Brador.

*Nobody here has great big refrigerators like we do back home, so a lot of stuff, like milk, is shelf stable. It's also why ice is so uncommon and why soda is served room temp. And why the Germans don't stock up—no space. So beer is made fresh and bought daily, like how the French consume their beloved bread.*

*I had a weird experience the other day that I'm still pissed about. I caught this guy staring at me in the shower after morning PT, and I asked him what the hell he was doing. He told me that his Southern Baptist minister had said that Jews were born with horns and a tail as punishment from God for being in league with the Devil, and that if you looked close enough you could see the scars from where they were cut off. Can you believe that shit?*

*When I said facetiously that I would pray for him, he told me not to bother because the Reverend Falwell also said that God doesn't answer the prayers of Jews. So then I told him to fuck off and I tell ya . . . that asshole better stay out of my face. Seriously, what the fuck?!!*

*All right, so do you have any of those wild and crazy college parties lined up for New Year's? All those young women so far from home—you are so fucking lucky! We'll be on standby guarding a nuke and probably will be into January. So have a glass of champagne and kiss a girl for me. Next year will be better 'cause it can't get any worse.*

*Well, this is a wrap. I'll see you in thirteen months.*

*Love Adam.*

# FOURTEEN

## Eiffel Shower

There is often a light rain falling around the Eiffel Tower in the month of December. Just remember to keep your head down and your mouth closed. This story will help explain why.

A couple of days before the dawn of 1985, my company was unexpectedly relieved from its "alert" status. My suspension had just been lifted, and it was one of the few times I was allowed to leave the country on my own. Pug and I each got a four-day pass, using up all our leave time to do so, and decided to catch a train to Paris without any arrangements whatsoever (beyond cocktails).

For the ride I decided to take piña coladas, using Stroh's 160-proof rum, along with Coco Lopez and pineapple juice, mixed and toted in a plastic one-gallon milk jug. By the time we got to Paris, it was dark and we were so shit-faced that we had to lean against each other for support as we staggered toward the center of town. The Citroen-clogged cobblestone streets are narrow, and the buildings crowding from all sides are so tall, that the city is made more of mazes than lights—more obfuscation than illumination.

We stopped by a small neighborhood grocery store, Middle

Eastern by the look of it, to ask for directions. *"Où est La Tour de Eiffel?"* I asked the guy behind the counter.

I had grown up with French but was far from proficient so I was keeping it simple. I figured anyone who lived in Paris would know where the Eiffel Tower was. We had to be close, but with the city planning as it was, we couldn't tell what was just around the next corner.

*"Je ne comprends pas,"* the unshaven grocery clerk said curtly, without even looking up. *I don't understand.*

I had used that phrase a lot in French class myself—mostly because I didn't have the confidence to speak in another language and I was obstinate. Growing up with a speech impediment lends one to the occasional lack of self-confidence when enunciating, so I said it again as clear as possible. *"Où est La Tour de Eiffel?"*

Again the same reply. I knew this guy was fucking with me because the only way he didn't understand what I was saying is if he didn't speak any French at all.

Looking him in the eye, I tested him with, *"Mange la merde et morte!"*

My French may not have been perfect, but this time my message was clearly understood: "Eat shit and die." In an instant, I received a pressurized stream of mace from a small can hidden in the bastard's fat hand. I was drunk enough to just stand there while he sprayed my face like he was putting out a fire. Then, from out of the back room emerged his twin wearing a bloody apron and waving a butcher knife.

Pug grabbed my shoulder and yanked me out of the shop so hard that I almost lost my shoes. The butcher was running out the doorway in hot pursuit when out of nowhere, a car smashed straight into the parked cars on the street next to us. We all jumped simultaneously, then stopped and stared at the crashed vehicles.

In that slightest pause, Pug and I made our getaway. Zig-zagging through the streets in fear of an insane knife-wielder, we finally came upon a parked police car.

Composing myself, I asked the officers, *"Où est La Tour de Eiffel?"*

At that point, Pug was full-on ready to run. He didn't speak French but he did know what happened the last time I uttered that phrase. Without looking up from his newspaper, the gendarme gestured lazily to his right. *"Là,"* was all he said. What is it with these French? We turned to follow his finger—and there it was.

From the grand terrace of the Chaillot Palace to the tower below are countless landings comprised of countless steps. I say countless because when you proceed to fall down each and every one of them, it seems to take forever and counting is quite impossible. Somehow Pug managed to stay on his feet and at the very bottom was kind enough to help pick me up. There before us loomed Gustave Eiffel's great erection—just a massive scaffold antenna really—but fascinating in a retentive mode.

The elevator that takes people to the top had a doorman who didn't like the looks of us from the start and refused to let us board. Maybe he watched our approach and had gotten the wrong idea, so I reproached him thinking I could reason with him. His well-practiced boot came swiftly to my crotch. I never saw it coming and it dropped me. My first instinct was retribution, but I was smart enough to know that where there's one uniformed guy there are more, and I didn't want to end up in the Bastille for vacation.

Again Pug was at my shoulder helping me up, and we went stumbling through the night until we reached a row of hotels. The first three told us there wasn't a single vacancy. It was, after all, the night before New Year's Eve. Same with the fourth. But in this one, I got an immediate sense of something, like drunken intuition, and told Pug to quickly follow me. Skirting the front desk, we took the wide staircase to the bottom-floor restaurant where all of the tables were covered in long cloths. A pair of locked glass doors loomed ahead, but something told me to check them anyway. Sure enough, the last person out had left one of the doors unlocked. Pug followed me in and to the back of the dining room, where we took refuge under the farthest tables and promptly fell deep asleep.

Just before five the next morning, the breakfast shift came in and started setting up the tables. When they got halfway across the room toward us, I crawled over to Pug who was still gently snoring and shook him by the shoulder. "Just do what I do," I said softly.

I stood up and Pug followed. We paused to straighten ourselves, and then we walked out of there like we had just finished a meal. The staff watched us in stunned silence, as if they were watching departing ghosts.

We strolled the chilly early-morning streets of Paris, sidestepping the homeless who sleep on the sidewalk grates for the warm air from the subway below. It was not the first time I had seen the world for what it truly was: grim and harsh, just like on the streets of every big city I have ever walked through in the wee-morning hours. It was a reflection of a planet that didn't care about us; no matter what we puny humans did, it was going to keep on spinning with indifference. That resignation was reflected in the faces of the crumbled bodies strewn about the age-old architecture.

In contrast, by nine that morning Pug and I felt that luck was with us, and we walked back to the hotel we had stowed away in. Turned out they had had a cancellation and we were able to book a room. It was the first time either of us had seen an "honor system" mini-bar and were admittedly shy on the concept. Many tiny liquor bottles and candy bars later, we were both passed out comfortably in our twin beds and slept through our first day in Paris.

That evening I awoke with a plan and a full bladder. I was still upset at the Eiffel Tower's elevator doorman and really wanted payback. Being sober and cleaned up made enough of a difference for the elevator guard—the same one from the night before—not to recognize us, and he let us go on up.

At the top of the tower is a small round observatory deck that encircles the structure's uppermost level. Recessed in the center of that deck, and barred by a locked chain, is a spiral staircase. Channeling Arthur Fonzarelli, I gave the chain's padlock the yank of its life—and it opened. Up and around the little circular stairs I went. At the very top, I found a small hatch door, less than half the size of a regular door. Its handle yielded to my grasp and it opened too.

I stepped out onto the thin metal grating that crowned the observatory, which was guarded by only a meager thigh-high wire rail. The wind whipped around me and a gentle rain hit at all angles. I felt like I was standing in a high and mighty crow's nest above all of Paris. Checking my bearings, making sure I was positioned directly across from the palace's dome, I was right above the offending doorman, some nine hundred and twenty feet below. From my vantage point looking down over the edge, I could just barely see the maroon speck that was the top of his hat. I hadn't relieved myself when I woke up and was nearly full to bursting. Doing a little dance to keep it together, I glanced down through the grates and into the observation deck. A swarm of camera lenses were looking up at me from the group of mostly Japanese tourists (which makes me wonder if I'm an obscene photo in someone's album of their Paris vacation). Without hesitation, I unzipped my pants and took a truly magnificent piss off the pinnacle of the Eiffel Tower.

I had Pug on the floor below to run interference if necessary, but no one tried to stop me. After I finished, I tucked myself back in and returned to the observation deck, then rode the elevator to the bottom. The Japanese took it all in with great humor. Several even congratulated me on the way down. At the bottom, I caught a glimpse of the doorman. There was visible precipitation on his hat and epaulette—and it wasn't from the drizzle.

Our last night, Pug and I reckoned that we had enough money left over to either have a really fancy Parisian meal or to get really drunk. We settled on the latter. But before hitting the bars, we needed to put something in our bellies. We filled up on burgers topped with stinky Swiss cheese at a little outdoor hamburger cart, which we washed down with a couple bottles of France's cheapest as we sat on a park bench by the Seine.

After that, we didn't feel so good.

Instead of going out that night, we went straight back to our hotel room. About four in the morning I heard Pug get up, stumble his way through the dark to the bathroom, and start vomiting like his insides were forcing their way out of him. It sounded painful, but as he shuffled back to his bed, I started to laugh.

"What's so funny?" Pug asked, a little hurt by my callousness.

"'Cause I'm next!" I blurted, jumping up and dashing for the toilet.

Until two hours before our train left we stayed in our room puking our guts out. Then we slowly made our way on foot to the train station, pale as sheets, leaving Paris the way we came: wobbly and leaning on each other for support.

That is my lasting memory of the City of Love, and it explains my enduring revulsion for hamburgers with Swiss cheese.

*Bonjour Harvey,*

    *Happy New Year's from the City of Love!*

    *The Paris scene is très international. At this bar last night, I was
sitting with people from three different continents. It's kind of a
happy result of the Parisians' snobbery. They don't want to mingle
with us foreigners, so the clubs are full of tourists from everywhere
else—a very cool global scene.*

    *I ran into a guy from Ottawa if you can fucking believe it. He
follows rock bands around Europe and sells homemade hacky sacks
at the concerts to stay afloat. I've got one in my pocket right now, so
I'm cool there. His next stop is London for Supertramp—what a way
to make a living!*

    *Got to soak up a little culture and went to the hallowed Harry's
(New York) Bar. It was a favorite hangout for the likes of Ernest
Hemingway and Jack Dempsey. I had some Armagnac older than
the both of us—good shit, man. Around here people smoke hash
crumbled up with tobacco and rolled, sometimes with a cigar
wrapper. This guy lit up a huge hash stogie right in the bar and I
think everyone got stoned. I tell you, tobacco gives me a meaner
buzz than hash, and I've got to stay away from it . . . it makes the
room spin.*

    *Pug and I played hacky sack under the Eiffel Tower on New
Year's Day. We also walked down to the Notre Dame Cathedral. It is
one scary old place. I could feel the ancient hatred, and a mean wind
was whispering "Get Out." Pug couldn't hear it. He's Catholic and I
think it's a Jewish thing. There is a definite undercurrent of anti-
Semitism here—always has been. But you'd never know it if you're
not one of us.*

*So many people think Paris is all fairy-tale romantic, but that's only when you're on the inside looking out. We saw this guy lying on a bench very early in the morning—all covered up with newspapers and snow. I told Pug he was dead for sure, but he didn't think so. On our way back he was being taken away by ambulance medics.*

*The last time we went to the field, a newbie froze to death in his sleeping bag. They called it "exposure," like how you "expose" water in the freezer to make ice cubes. I call it lack of training and $35,000 worth of cold comfort to his family. Thank God for my Canadian anti-freeze.*

*Hey, why did the French plant trees along the Champs-Élysées? Because they knew the Nazis liked to march in the shade!*

*I met this French chick at the disco. At first she wouldn't give me the time of day, but after a few drinks she agreed to dance with me. At the end of the night, she took me home with her. The thing about banging a French girl, beyond the paper thin walls, is when she started to orgasm, she shouted "Yes, yes, yes" over and over. Except she's French so she was saying "Oui, oui, oui." After a while it sounded a like a pig squealing. And the second that's in your mind, you're in trouble because you can't laugh and keep a hard-on at the same time. Literally, too fucking funny!*

*Wait . . . I think I can hear Edith Piaf in the distance. Write me back.*

*Love, Adam.*

# FIFTEEN

꞊ꞏ꞊

## Drop and Roll

I t's true that the Third Armored Division was not engaged in combat. But, propelled forward by the Reagan Doctrine, we were right on the precipice—shaking our spears and rattling the Soviet Bear's cage. The aftermath of the Abel Archer War Scare of 1983, the mock invasion of the USSR by a combined NATO force, is when the Cold War instantly got red hot. The Russians didn't believe that it was merely a war game the US-led coalition was playing, and in light of President Reagan's rhetoric, considered the possibility of war very real.

It was widely believed that whichever superpower fired off their missiles first would stand the better chance of survival amid the chaos of a world laid to waste. To the Soviets the Able Archer exercise looked like America was preparing for the first strike and the Communists went on high alert. The USSR's only chance was a quicker and more decisive preemptive nuclear strike. While orders were awaited from the Kremlin, nuclear missile launch switches were nervously fingered, and the massive Soviet army began to muster all along their border, ready to repel the onslaught. The implied threat of those military maneuvers brought the world closer to a full-blown nuclear world war than ever before—

and disaster was only narrowly averted. Ronald Reagan's bluff had very nearly backfired horribly.

Two years later, in January of '85, Reagan was not ready to let up, and the US armed forces joined NATO troops in a Return of Forces to Germany exercise known as a REFORGER. Nicknamed "Central Guardian," the colossal military maneuvers utilized a combined multi-national fighting machine of almost a hundred thousand soldiers. What we were doing was more than just a mock-up—we were running the actual game plan designed for an attack on the Soviet Union. To the Russians, it once again looked like a capitalist occupation was imminent, and they reacted predictably.

Our mission was to run the Soviet Army into the ground by forcing them to shadow us as we performed our battle-ready postures along their border. Reagan's plan was a war of financial attrition, designed to bankrupt the USSR no matter the cost.

It's not cheap to keep an entire army fed, supplied, and on the move. "You guys don't know how good you got it," our first sergeant told us during a morning formation. "The Red Army grunts are down to one meal ration a day. Most of those guys don't even have underwear, and they have to use rags wrapped around their feet for socks."

We were skeptical when we heard this; it sounded too much like P.R. But twenty years later it was confirmed to me by a Soviet veteran driving a taxicab in Chicago. Turns out, we really did have our Russian counterparts running on fumes.

But while we were breaking the back of the "Evil Empire," the American GIs were being run ragged as well. Being in the field meant living a harsh nomadic lifestyle where we carried everything we needed on our backs, and all of our belongings were always packed up and ready to go. On a daily basis we pulled up and set down stakes at all hours of day and night, working in sub-zero conditions and sleeping in frigid canvas tents. Our meals were eaten standing up outside with burning-cold fingers just yards away from the mess truck, huddled in small groups like penguins. Nothing was simple or comfortable. Just going to the bathroom involved walking for twenty minutes through the frozen woods so you could find a private place to dig a hole in the snow to shit in.

In spite of all our cold-weather gear, life under those conditions was one of constant exposure to the elements. Lips and fingertips would spontaneously split and crack open, and hypothermia and frostbite ran rampant, like everyone's noses. Our unbreathable heavy rubber "Mickey Mouse" boots we wore caused terrible cases of trench foot. The magnitude of the event had our resources stretched thin, but in a real-world scenario we were the disposable first wave—not meant to survive, let alone be in comfort. And our greatest enemy, the absolute life-sucking cold, stalked us relentlessly.

My company, the Death Dealers, had just pulled into the thickest part of the callous that separates Germany from the Czech Republic. Only a small part of the Spearhead Division, we were comprised of a whole fleet of hulking howitzers with their matching APCs, five-ton supply trucks, chow-wagon, water buffalo, and a whole fleet of light trucks and officers' jeeps. Quite the entourage. Even when a little bit of the Army moves, it moves big.

One bone-chilling afternoon, just as we arrived at our coordinates, we were put on standby. There was no sense in setting up shop until we were sure the brass monkeys hadn't screwed up and put us in the wrong place. During maneuvers it was important to be properly coordinated with our foreign counterparts. We wouldn't want our armored vehicles overrunning somebody else's infantry, and that cooperation takes a lot longer when dealing with combatants from other countries.

So we were all standing around freezing our asses off for so long that the higher-ups realized they needed to keep us from freezing solid and sent a couple sergeants, with a squad of enlisted, to set up a tent with a heater. When you're in the field, life is a game of checks and minuses, and everything you do to put yourself in the hole accumulates. The constant hunger, exposure, and days on end of sleep deprivation all take their toll. Staying alive didn't mean being able to pay the bills. Existence got stripped down to the barest essentials, and every single day was a struggle with mortality.

That day, it was taking too long to get the heater going, and recognizing that survival was not a passive state of being, I left the huddled masses and made for where the parked vehicles were. Sharking through the rows of unattended trucks, I looked for the glint of keys through frosted windows. Anything I could do to put the balance in my favor was another step toward walking out of that minefield on my own—*walking* being the operative word.

A few weeks prior, when we left the post for the field, I had gotten clearance from the communications lieutenant to ride in their enclosed APC. I was dressed accordingly, as it gets warm in those things. But at the last minute, Sgt. DumbOx ordered me into the back of an empty five-ton truck, without any of my gear, where I sat by myself in subzero conditions for over six hours. By the time we pulled into position, I could no longer feel my legs from the knees down. Later that night, the feeling returned to everything but my toes, which stayed hard, cold, and unresponsive, like little rocks in my boots. I had the distinct impression that when I took off my socks, my toes were going with them.

I had a fantasy of moving to California after I got out of the service. The thought of someplace where I could be warm and wear shorts year round was intoxicating, but what was I going to do without toes? I pictured myself standing on the beach trying to bury my feet in the sand so no one would notice. "What kind of woman is going to want a toeless man?" I thought. After two weeks of rubbing my feet every chance I could get, my toes slowly started to come back to life. But they now ached intensely in the cold, so staying warm had taken on a whole new dimension.

The instant I spied the truck with a set of keys dangling from the ignition, I returned for Pug and, as inconspicuously as possible, we separated from the group and double-timed it back to the oasis I had found. Protocol be damned, at least we'd be warm—even if we did get in trouble. I jumped into the cab, locked my door, and started it up. Pug climbed into the other side, and once I got the truck idling we slid down out of sight. The hot air from the heater felt like a gift from God as it drove the chill from my limbs. I could feel my chi resurging, though the glorious feeling was short lived. No sooner was the cab warming up than Pug's unlocked passenger-side door was yanked open.

It was Sergeant Shrivel. A paragon of mediocrity, he had been in the Army long enough to be a sergeant major but was not motivated or competent to be given a real position of power. Close to retirement, Shrivel concentrated on sharing the misery of his life with all those he came in contact with, and he was respected accordingly. Shrivel had seen our truck's exhaust and barged his way into the cab, pushing Pug into the middle of the seat so he could close the door.

For a moment, we thought he just wanted to escape the cold too, until he barked, "Get the hell outta the truck."

He wasn't standing outside the truck, holding the door open, telling us to get out. He was trying to kick us out into the cold while his sorry ass stayed inside, taking advantage of the prize I had found. We not only refused to budge, we ignored him completely while his mouth ran off. We were "wasting diesel," he said repeatedly, then ordered us "to get all the enlisted into the warming tent." What a crock of shit. The tent he was talking about was set up just fifty feet from the truck we were sitting in and it didn't look warm to me. I could clearly see the tent's opening from my seat, and nobody was going inside. And if the tent offered warmth, why was Shrivel staying in the truck while trying to kick us out? Pug was acting as insulation between us, and as much as Shrivel tried to get my attention, I refused to even turn my head his way. Even if we were only delaying the inevitable, we were gaining precious seconds of warmth—and every little bit mattered.

"Look at me, damnit," Shrivel roared. "I told you to get out of this fucking truck and go into the tent—NOW!"

Still we ignored him as if he wasn't there. In the truck I was warm; the tent was more likely a hollow promise—like everything else in the Army.

Then, in an atypical shift, Shrivel resorted to reason, assuring us that Sgt. Brown and Sgt. Maldonado already had the kerosene-fed heater going. But Pug and I continued our game of pretending we couldn't see or hear him, and that really pushed the paunchy sergeant's buttons.

Frustrated to no end, Shrivel finally gave us the ultimatum to "get the hell outta the fucking truck and into the fucking tent—and that's a direct fucking order!" Of course he made no attempt to get out of the truck himself.

Glancing again at the tent, I could clearly see there was no smoke coming from the stovepipe poking out the top, and I was convinced Shrivel was lying. When people are forced to live outside in harsh conditions they become brutish. The learned humanity recedes and the inherent animal comes out. You become dulled to your own pain and insensitive to that of others. That's when civility falls by the wayside and getting fierce happens fast. To me it was all about self-preservation—orders be damned. All I cared about in that moment was staying out of the cold, so I continued to stare straight ahead and refused to move.

Sgt. Shrivel finally got fed up with me blatantly ignoring him and reached for the keys. Before he could pull them from the ignition, I had his arm by the wrist with one hand and just above the elbow with the other—and I started twisting, hard. I shouldn't have done that; an enlisted putting his hands on a sergeant was deeply frowned upon, and I knew in that moment I had just fucked up. But my switch had been flipped, and I was going to teach that piece of shit something about respecting another's personal space.

As I was threatening to dislocate the sergeant's shoulder, with Pug caught between us, divine intervention stepped in. Sergeants Brown and Maldonado came streaking out of the tent in front of us with their heads fully engulfed in flames, zigzagging through the snow like human torches.

"Drop 'n' roll, drop 'n' roll," I shouted at them through the windshield.

I doubted they could hear me, but as if on command, they both dove head first into a snow bank. The stunned silence from Shrivel was priceless as the two sergeants tossed about in the snow with a synchronicity that looked rehearsed, rolling on their backs and kicking up their heels as if they were still running.

Smugly, I finally spoke. "Do you still want us to get in that fucking tent?"

The sergeant didn't answer, only pursed his lips so hard I thought they'd crack.

Instead of being horrified by what we saw, it struck both Pug and me as so funny that we broke out laughing until tears

streamed down our cheeks. That bizarre scene was the most entertainment we had had in weeks of living in the field. We started
imitating how they were rolling in the snow, rocking the cab of
that big truck. Sporadically one of us would shout "drop 'n' roll!"
and we'd lose it all over again. I think I peed myself I was laughing
so hard, and the longer we laughed, the angrier Sgt. Shrivel got—
which fueled our antics all the more.

When the singed sergeants sat up, the enlisted guys, who had
been helping them set up the heater, heaped piles of snow onto
their heads like Elizabethan wigs. It was the most ridiculous spec-
tacle I had ever seen, especially in that time and place. As Pug
and I continued cracking up and imitating the sergeants on fire,
Shrivel sighed loudly and let himself out.

As inappropriate as it may have been, I don't think I laughed
so hard my entire time in the service. The unrestrained humor of
the moment was amazingly cathartic, almost wiping away all
those days of living in misery.

Once Shrivel was gone, I turned the truck back on to warm it
up again—and Pug locked his door this time. Then, like two buddies
on a couch, we sat back to watch the rest of the show in comfort.
All we were missing was a big bowl of popcorn perched between us.

For the record, Sergeants Brown and Maldonado received
purely superficial burns thanks to the extreme cold. A medic ap-
plied thick salve to each of their faces, and they returned to their
units in the field that same afternoon. After the layer of burnt skin
on their faces died and dried, it turned into a gray death mask.
Eventually it sloughed off, leaving both sergeants looking years
younger and a lot rosier (though they both had a constant alarmed
look about them until their eyebrows and lashes grew back).

A word of caution: Do not attempt this type of "chemical"
peel at home and expect the same results.

Hi Harvey,

How are you? I heard Grandma's not doing well. I feel terrible saying this but I think it's her time to go. She's had a full, rich life and she's been sick for a long while. It's like Prince says: "Parties weren't meant to last." Still, it's a real shame to see the old lady give up the ghost. As grandmas go, she's definitely a good one!

But you can't fear the reaper. Death is as much a part of life as life is a part of death. I'd like to think that death is just a transition phase, like a chrysalis before our true form can be realized . . . or maybe it's the eternal piña colada on the perfect beach. I'll let you know when I get there—isn't that what Houdini said?

I'm surviving right now but not having that much fun doing it. Have you heard anything on the news about the large-scale military training maneuvers here in Germany? For the last two weeks we have been fighting an imaginary war and I think they're winning. Does America know what American boys and girls are doing over here? I've been cold, wet, and miserable all day, every day. The weather is horrible—freezing rain and knee-deep slush—and my boss is a complete asshole. Maybe it's just me and my inherent lack of respect for authority, but my gun sergeant DumbOx and I don't get along at all.

Oh look, Kash just came into the tent and handed me a beer. It's afternoon and we're in the middle of an onion field near the Czech border. Not exactly by the rulebook, but when you're eating disgusting dehydrated food, shitting in the woods like an animal, and sleeping in a smelly little tent, a modicum of civility like a beer, or a cigar, go a long way.

Did I tell you about the drug/alcohol program I'm being forced to take? It's run by this crazy ex-alcoholic officer's wife, and she's constantly on my ass because I'm not going along with the program. It's all because of my solid criminal conviction, which is brilliant considering there was no evidence. Just one gram of hash in somebody else's locker. And you wonder why I'm drinking beer in the field in the middle of the afternoon—medicinal porpoises only, I assure.

But the bottom line is, if you don't actively stimulate your sense of humanity, it can slip away so easily. And it's the little things, like getting a letter from home that reminds you of who and what you are. I haven't gotten anything for a while, though. Maybe I should threaten the mailman . . . do you think he's holding out on me?

Hey, can you send me some acid? Three hits if it's good shit, six if it's just so-so. Me and a couple friends have been talking about it and we want to take a little trip. Oh, and don't worry, they don't test us for it. And trust me, I'll only do it when I'm off duty and away from the weaponry, I promise. You can just put it in a book or magazine between the pages. Cool? Cool!

I really need to get the fuck out of here, even if it's just in my own head. I want to forget what it feels like to be an indentured servant for a little while. It's funny how you lose your own freedom when your job is to guard the "Land of the Free."

Well, sleep tight cause I'm still on that wall, but I'm coming down and I'll see you soon.

Love, Adam.

# SIXTEEN

## Wrong Turn at Edelweiss

I was on restriction during most of my time in Germany for one infraction or another and wasn't allowed out of the country. Hell, more often than not I wasn't even allowed off post. The window of opportunity that let me go to Paris was the brief exception. But when our battery went, en masse, to Innsbruck, Austria, I had to be included whether I wanted to be or not. How else were they going to keep an eye on me? I asked my top sergeant to let me stay behind, but he only laughed as he kicked me out of his office. The ski trip was in appreciation for surviving REFORGER and was compulsory—like an unwanted gift that couldn't be refused.

I had sworn off skiing years ago, and as a Jew knew too much about Austria to ever want to visit. But I had no choice, and that really rubbed me the wrong way. Besides, we had just spent a month living in a frozen hell. Why the fuck would I want to go play in the snow? The expression "misery loves company" sprang to mind, and I decided to help my top sergeant regret forcing me to join in.

Because the spouses of the sergeants and officers were encouraged to come along, decorum was, literally, the order of the day. Specifically that meant all of us lowly privates had to behave ourselves, which immediately invalidated the trip as a vacation. I didn't care how loose the yoke was around my neck—I still resented the restraint.

My entire Alpha Company all got loaded up into a great big touring bus with the officers up front, the privates crammed in the back, and the sergeants, as always, in the middle. I cued up Pug's boom box with Jim Morrison's rant about petitioning the lord with prayer. It seemed proper to let the upper echelon in front know that pleas to the Eternal would fall on deaf ears. If the lower enlisted rabble I was part of had the privilege of drinking alcohol while sitting in comfort for six hours because the activity was labeled as "R&R," we were going to turn it into a party. Privates were the hardest worked and most expendable cannon fodder there was. We lived like every day could be our last. And this was our chance for the tail to wag the dog.

Like most, I had started off in the service of Uncle Sam naive and optimistic. But that pendulum had swung far the other way. I was tired of being screwed over by the Army's "big green weenie" and being looked down upon as an inferior by schmucks who had more rank. So I seized on the opportunity to get under their skins, replaying Morrison's tirade over and over until groans of "how long is this going to last?" came from the front of the bus.

The guys around me were more amused than anything else; it was always entertaining when one of us got uppity and pushed the limits of our cage. To those elitists up front, we were third-class humans at best, a nuisance otherwise, and unfit to rub shoulders with. For them to even make casual conversation with an enlisted was considered fraternization and strictly prohibited.

"Fuck 'em," I thought. "That's what they get for being in my army. How do you like me now?"

In the seat right in front of me was the top sergeant's snitch and favorite pet, the Fat Finn. Finn was of Norwegian descent, with a platinum blond crew cut and tiny pale blue eyes recessed into his doughboy face like glinting raisins. To say he was big-boned would have been modest. Dinosaur bones was more like it—not to mention he was covered in a solid layer of blubber that was enough for him to live off for a month.

Beyond being Top's personal secretary and valet, Finn also acted as the enforcer of the rules amid the enlisted. At best, he was a small-minded yes-man in a bouncer's body whose side job was to make sure we didn't get out of hand. The top sergeant had set Finn on Pug and me like a sheepdog, and watching us fuck-ups was a detail he deeply resented.

As the top sergeant's bitch, Finn enjoyed a privileged status, but it wasn't without sacrifices. For example, Finn had realized that to make the perfect pot of coffee for the top sergeant each morning, he had to forego his morning PT. At least that's what he claimed to be doing every day while we were out running in the crisp early light as part of our physical training. Finn also liked to hang out with the upper crust and rub elbows with the officers, pretending he was one of them, though they deemed him nothing more than a busboy and server. Finn had the aspirations of all proper bootlickers—to one day have others grovel before him.

The very first morning at the chalet, in typical Army fashion, the entire entourage hit the slopes of fabled Innsbruck just after six a.m. It was a little before daybreak, but we could see the rising sun in the distance. The hills surrounding us promised to either be a skier's paradise or an amateur's snowbound hell.

I had skied before and was already quite adept at falling, but Pug had no experience whatsoever. So we decided to hit the "kinder" slopes first, where I could impart all my skiing knowledge to Pug. Everyone else was headed to the intermediate hills, but pride goes before a major wipeout, and for safety's sake Pug and I were going to be humble about the endeavor. The directional signs had no English translation, but I figured we could manage.

We started by choosing an uncrowded ski lift and off we went. It was odd that no one else got on that lift, but we figured it was because we were the only amateurs around; it looked like it was going to drop us at the top of the smallest hill right in front of us. But when we reached that little hill, the lift kept going, like it forgot to stop. The next station was only a faint snowy shadow in the distance, but as we got closer, we started to get nervous. Were we on the right lift? When it didn't stop there either, I pulled out the joint I had rolled. As we continued to climb farther and farther from where we started, there appeared to be no end point. After forty-five minutes on that precarious dangling bench, stoned and frozen, we finally got to the end of the line. The dropoff was completely deserted and Pug didn't want to leave his seat.

"Can't we just stay on the lift and ride all the way back down?" he said.

I couldn't blame him; it appeared we were being dropped off on the very top of the mountain—which was no more than thirty feet across. But I made him get off with me, insisting we could ski our way down, though I truthfully had no idea how far that was.

I jumped off and Pug followed. We carefully sidestepped to the closest edge and looked down. As far as we could see, it was a straight drop to a bottom obscured by clouds below. We then slowly sidestepped to the other side. From there, the mountainside appeared to be concave. With no approach that was remotely skiable, we were stuck on top of an alp—in a space no bigger than a small office.

How the hell were we going to get down?

Just as our plight seemed hopeless, the sun's rays struck the top of the alp. Suddenly the mountain stepped out of the shadows and we realized we were standing on the edge of the world. The snow around us glistened like crushed glass. Then, like a messenger from the gods, along came the only other human we would see all day. As the light hit us at a sharp angle, he dropped off the lift right in front of us. But before I could plead for him to go get help, he skied away as if we weren't even there. Pug and I looked at each other wide-eyed. We didn't think it was possible but we had no choice. Feeling like we would plummet to our deaths at any moment, we valiantly tried to follow in the trailblazer's ski tracks.

I was able to stay on my skis for almost fifty feet at a time, which was pretty good because each time I fell, I tumbled for more than a hundred yards. Pug was not so adept and spent every few yards being violently thrown to the ground by gravity, which takes a lot out of a person. After a couple hours on that slope light years beyond our skill sets, the trail we were on abruptly ended and became a cliff. In the middle of that pristine wilderness, on a towering metal post, stood an indecipherable metal sign.

I held on to that post for balance for a good twenty minutes before Pug came into sight, skiing in the best form I saw him in all day: one ski down, the other airborne while he leaned back and wildly flapped both arms. As he approached me, I tried to warn him that the trail ended, but he shot right past me like a rocket and disappeared. I spun around, but he was gone.

Then, out of the cold blue air, I heard a thin voice. "Adam?"

"Pug?" I answered back. "Where are you?"

"I'm down here," said Pug's distressed voice.

Holding on to the signpost with one hand, I leaned way over the edge. It was a sheer drop of about two hundred feet to a small run-off stream full of large boulders. About six feet down the side of the cliff, an old tree had lived and died out of sight, and Pug had got himself caught up in its skeleton.

"Do you remember when I told you to listen to me very carefully or you could die?" I called down.

"Yes" came the feeble reply.

"Throw your skis and poles up to me as hard as you can, okay? If they don't clear the top and fall all the way back down there, you'll have to go get them. Okay?"

His voice was shaky but he managed an "Okay," followed by four strong hurls of his gear up over my head. Clearly his adrenalin was kicking in.

I held on to the post with one hand and extended my ski pole to Pug with the other.

"Take hold of this," I told him. "You're going to have to let go of the tree."

Pug hesitated, then grabbed the end of the pole. I hoisted him up as if my own life depended on it. After he landed next to me, we took a long minute to catch our breath, lying on our backs, side by side, staring up at the sign.

Finally, I said, "I guess that sign means 'Look out for the cliff!'"

Pug gave a half-laugh. "Yeah, maybe we better memorize the wording in case we see another one."

We backtracked a bit and came upon a small hairpin turn that led to a narrow pathway. It seemed to go from one range to another, but it was heading our way, downward, and we took it. We then made our way from one alp to the next, across what must have been a temporary snow bridge. That was our big mistake, though I don't know what other choice we had. We had no option but to ski and fall and fall and ski till well after the sun went down, never seeing another sign of life. For more than fourteen hours, subsiding solely on virgin snow, we carved a slapstick path through the wintery solitude of that never-ending hill.

Thankfully it was a bright night. At no time did I ever consider the situation life-threatening, but I knew we'd have to either find shelter before midnight or dig a snow cave and spend the night where we were. We were both hungry and tired, and we'd been in the cold for far too long.

I was staying ahead of Pug by just a few minutes when BOOM. I skied right into the only tree in the center of a small valley, smacking my head on the trunk. I fell back into the snow and started to laugh. I always thought skiing was foolish, and now it was the skis that were turning the tables on me. The irony was relentless.

Lying on my back, I saw Pug coming down the hill, still falling every ten yards or so, and viewing him from upside down made me laugh even harder. Pug was fed up with the whole ordeal, and he was even more fed up when he skied right into the same tree I was still lying at the base of. He was going faster than I was, though, and struck the tree with more force than I did.

When he caught his breath, he asked, "Why were you laughing?"

"Because I knew you were going to hit this tree too!"

Pug failed to see the humor in it.

The big hunter's moon shone brightly on a silvery strip ahead of us in the distance, and we knew it had to be a road. Just a couple more miles of misery before we reached a link to civilization. When we finally made it to the road, Pug, who had grown to hate those skis, took his off and started walking. The road ran gently downhill and I was finally getting the hang of them on flat ground. So I kept mine on, grinding them down as we went.

The road was gravelly, unpaved, and mostly iced over, and we hadn't seen a car yet, but our courage was renewed. It had to lead somewhere. We could see snowy little farmhouses in the distance, nestled up closer to the mountain, and I kept thinking about the warm hearths on the other end of their smoking chimneys. If it came to it as a last resort, we agreed to leave the road and heave ourselves through deep snow to plead at the door of a farmstead for refuge.

Close to eleven that night we came into a small darkened town, a haven in the mountain pass, and found the only open tavern. We were utterly exhausted and looked as much, but the sounds of human voices and the clinking of glassware were music to our grateful ears. The atmosphere seemed different than in Innsbruck; they served Moretti instead of Spaten and grappa instead of schnapps. We each gulped down a big bowl of minestrone soup with a thick wedge of dense brown bread. Afterward, we asked the proprietor to call us a taxi.

We both fell asleep on the drive back to the Innsbruck chalet, so I don't remember any of it. What I do remember is that the fare cost over two hundred schillings, including a generous tip, which was most of the money we had brought for the whole trip. It took all day and half the night, but Pug and I had skied down the wrong side of the alp—a feat no reasonable person would believe.

*Dear Handsome Harvey,*

*Well, if you receive this letter, you'll know I survived another REFORGER. Hope you can read this as I'm writing fast. I've got to be on a bus in an hour for an R&R trip to neighbouring Austria and I want to drop this in the mail slot before we depart. This is some bullshit put together by one of the officers who likes to ski and no matter how hard I tried I just couldn't get out of. But you know the deal, privates ain't got no rights.*

*As much as sleeping and shitting in the woods in winter sucks, I must admit what we did was pretty impressive. As far as I know it was the largest military maneuver in history! A multi-national force all banded together to repel the Soviet threat—although I didn't see one Russkie the whole time. But I did meet some guys from the Italian and French armies. It's funny that Italy started off as rulers of the world through military might. But now as a country they're old and retired, leaving the fighting to us. And like old people, they focus on the important things in life.*

*The Italian troops have these way cool uniforms, very snappy, and their MREs look homemade. They get chocolate, coffee, and a selection of liqueurs with their meals. But it's not all fun and games —they have a job to do too. They wear these thick wool socks and only every other week do they take them off and clean between their toes. Bet you didn't know that's where Parmesan comes from!*

*The French guys have the best rations; they look like they're catered from a chic restaurant. Very fancy . . . and they even get red wine. They laughed at my MREs and used a word that doesn't even exist in English: "epouvantable." I always liked that word. It means* the grossest of the gross. I must admit that French can be so wonderfully expressive.

I like how our officers refer to REFORGER as a "war game." It's the same twisted connotation as in big "game" hunter. It's only a game for those on top, the shot callers. Seems like every time we go to the field someone dies, and it's usually one of us guys at the bottom. I've got a new one to add to my list. This dumbass from the next battery crawled under his howitzer to get some shut-eye out of sight of his sergeant. When they got the call to move out, they had to turn around and pulled a 180° with the track vehicle. The guy napping underneath got ground up like sausage! I mean—just the wurst.

Okay, it's a bad joke for a really shitty thing. I tell the new guys someone's going to die—just don't be stupid. Don't let it be you. Sometimes, however, stupidity is unavoidable.

Hey, how's Grandma doing? You know you're her favorite. Is she still in the hospital? I'd think if this is the end she should be allowed to die at home in her apartment. I think that's what everybody deserves, not in some cold, sterile hospital environment. That's not where you want to have your last fleeting sense of humanity. When my time comes I want to be at home with my loved ones, dogs and family—in that order! And I want a big party with the soundtrack "Wish You Were Here" by Pink Floyd. Keep in touch and write when you can. I don't think they'll let me go home for the funeral.

Well, enough for now. I've got to show some Austrian women my horns and tail. Take care.

Love, Adam.

# SEVENTEEN

~~

## The Bull Finn

Nineteen hours after we had started out, Pug and I finally shambled into the Innsbrook chalet, far worse for the wear— but wiser for it. A bunch of the guys were still up and sitting around the fireplace drinking with Fat Finn brooding dead center. Wind, sun, and snow burnt, we looked like we had just returned from a Shackleton expedition.

"Look what the cat dragged in!" Finn boomed toward us. "Were you two blowing each other while the rest of us were out skiing?"

Finn was pretty tanked up and louder than he should have been; he was actually waiting up for us on orders. Our absence had been noted early on, and Top didn't want his vacation "fucked up beyond all recognition" just because Pug and I had done something stupid and gotten ourselves killed—or arrested. That's why Finn had been told to keep an eye on us in the first place. But since we had disappeared right off the bat and no one had seen us all day, Top was not happy, and Fat Finn was already in hot water.

"What do you mean you can't find them?" Top had barked at

Finn earlier that evening. The sun had set and everyone else was accounted for. Finn had no answers.

As well as being Top's pet, Finn was also the billet's bouncer. His little office was right next to the top sergeant's, connected by a back door, and if an unruly soldier went into Top's office unannounced, Fat Finn would throw him out. With more than two hundred sixty pounds generously packed on a six-foot-two frame, Finn definitely knew how to throw his weight around, especially among the lower enlisted. But Pug was the same rank as Finn, a specialist, and I was a hard-core private. Neither one of us was impressed—nor was I in the mood for his bullshit.

Pug was quick to remind me that we had reefer and beer back in our hotel room. That was the one and only time I saw marijuana, as opposed to hash, while I was overseas, and it was enough of a treat to distract me. When we got to our room Pug decided to lie down, so I showered first and got dressed. I still wanted to go get a drink and maybe get in that fat bastard's face. But as I sat on the edge of my bed to roll a joint, the whole world suddenly felt like it was caught in a spin cycle. Vertigo hit me hard, and I didn't have the energy to fight it off. I collapsed and fell fast asleep before my shoes hit the floor—fully clothed, with the lights on.

About three in the morning I awoke to a pounding on our door. Pug opened it up and Fat Finn poured in like a landslide. He was all liquored up and had remembered why he was so mad at us.

"Where . . ." he stammered, "where did you guys go?"

Finn then lunged toward Pug, who weaved in time for the wall to hit Finn hard, which made him even angrier. That's when I noticed the marijuana I had left out on the side table—seeded, with rolling papers at the ready. Finn was widely regarded as a snitch, and anything he saw or heard would be passed on to the top sergeant as political credit. He wanted desperately to rise up out of the common rank and file to make it into the Sergeant's Club, and it didn't matter who he stepped on to get there. In a panic I realized I had to distract him, preferably getting him out of the room altogether. But while Finn was full of bluster, he was also big and scary, and I had serious reservations about taking him on.

He fixed his blurred gaze on me. "Fuck you, Harris."

He was ready to charge and I realized I could either try to calm him down or really piss him off and see what happened. So when I said, "Fuck you, you're just Top's Bottom-Bitch," I knew I was taking my life into my hands. I had seen Finn in action. He came on like a raging bull, so I teased him like one.

As he was processing that insult, I commanded him to "Get me my pipe and slippers, you albino fat fuck."

His flushed face got rounder and redder. Finn had a short fuse and it was now fully lit. He jumped at me and landed on my bed as I dove out of the way.

"You sorry fat asshole" I continued, moving over to the doorway as he got up. "You can't even keep up in PT, which is why you make your daddy's coffee."

He came at me and was surprisingly fast. I ran out of the room just inches ahead of his hands as he chased me down the hall. Catching sight of his room's open door, I ran inside and slammed it shut behind me. Sometimes you've got to think on the fly and take one thing at a time. I had gotten him out of my room—but what was I going to do with the enraged Finn now? It was then that I was reminded of Martinez.

While I was laid up in bed with my crushed foot, I'd had the company of Specialist Martinez. He had gotten a large tattoo to celebrate his upcoming transfer stateside, and it had become badly infected. His punishment was to stay in bed all day, on a diet of antibiotics, while a butterbar ran around taking care of his paperwork. That way they each learned something.

Martinez was the tough guy of his crew and had only granted me respect after he watched me take on a big Marine in Frankfurt. I was an unpredictable brawler, but Martinez was a technical fighter. To pass the time one day, I asked him what the secret was to throwing a good punch.

"Don't close your eyes," he said.

"What?" I asked. *Could it be that simple?*

"Don't close your eyes," he repeated. "Most people shut their eyes when they throw a punch. It's the body's reflexive way of protecting them. Don't do it."

Martinez was a man of few words. I appreciated his candor and took it to heart. Realizing how flimsy the hotel room doors were, and not wanting to get pancaked beneath one by Fat Finn, I let go of the doorknob and stepped way back. When Finn slammed into his own door, there would be no opposing force. I was standing toward the back of the room, remembering Martinez's advice and trying not to freak out. It's a lot easier to be a berserker than a "Cool Hand Luke."

As Finn blew through his door like it was made of balsa wood, I steadied myself. Keeping my composure while Finn barrelled down on me was harder than it sounded. Finn outweighed me by almost a hundred pounds; he was belligerently drunk and I had pushed all his buttons. But for this to work, he had to be completely pissed off, and if my one shot wasn't perfect, it was going to make Finn all the more dangerous. I was scared as hell, but I stared down the charging Finn and, just before he violently enveloped me, I threw my punch—eyes wide open.

I struck that behemoth just off center of his big fat face and he dropped like a tranquilized rhino.

Pug appeared in the doorway and his look went from wild concern to amused amazement. There I stood unscathed with Fat Finn laid out in front of me. We threw a blanket over him, closed his busted door the best we could, and went back to bed.

The next morning Finn woke up with a lump like a potato on his forehead from the fall and a shiner around one eye from me. Best of all, he couldn't remember a thing. There were plenty of willing witnesses to his mass consumption, though, and with a broken door on our commanding officer's hotel bill, Finn ended up on the top sergeant's shit list—while Pug and I went unnoticed.

*Hey Harvey,*

*What's going down? I woke up this morning with a hangover, had a beer for breakfast, and damn if I'm not drunk again! This German beer has all the protein of a steak and all the carbohydrate of a potato—it's a meal in a can!*

*I now know where I want to go to school. Have you seen this month's issue of Playboy? You've got to check out the girls of Texas—holy shit!*

*Hey . . . do you remember the last time we went skiing? It was in Ottawa with the Bernsteins and I had that spectacular wipeout all the way down the hill, head over heels, and bent both my poles. And the people on the chairlift started clapping for me when I stood up. Remember I swore after that I'd never go skiing again? Well, I should have kept that promise.*

*It's not probable but it is possible to ski from Innsbruck right out of Austria as long as you take a wrong turn and keep on going. We got on the lift just as the sun was coming up and didn't get to the bottom of that first hill till eight o'clock at night—me and my buddy Pug. And while I can't say for sure that I've been to Italy ('cause I didn't get my passport stamped), I did have my first Moretti, a fine Italian beer. Not as good as a Canadian lager, but close enough.*

*And I learned a couple lessons from my ski trip: 1) never go on vacation with your boss; 2) never take a tour—unless you're in charge; 3) a vacation should not require effort or threaten your life— that's called work. Oy!*

*On our way back to Hanau we stopped off at a souvenir shop in a town known for its crystal, and people were throwing down serious ducats for delicate shit that's just going to get broken in transit. One of my stupid sergeants bought a chandelier! I got a little round rough-cut ball of quartz that looks like a cloudy marble. It fits in the pocket easy and with just a little polishing would look really cool. It makes me think of the old blues refrain: I might be a lump of coal—but I'll be a diamond some day.*

*Diamonds are only created under extreme pressure. So when the weight of the world upon your shoulders starts to crush you down, just remember—it's a refining process. And dammit . . . I am clearly still in the lump of coal stage.*

*There's an old Korean saying I like: "Adversity is the mirror that reflects one's true self." It's what the restaurant owner said to me and my buddy when he wanted us to buy rice from the PX for him. Man, that was some good bulgogi—and all the sake we could drink! Oh, glory days . . .*

*I'm getting thirsty just thinking about it so I think I'll have another beer. Ah, there's nothing like getting boozed up early. I'll probably be asleep by three. Hope you have a fun and productive day too.*

*Love, Adam.*

# EIGHTEEN

---

## Tanks, Louie

When you're in field artillery, you're always in the field—somewhere. And somewhere along the northern German/Czech border, as part of a large-scale military exercise, I ended up driving one of our armoured personnel carriers with Sgt. Louie late one night. Louie was the sergeant's first name, and it wasn't important enough to me to know his last. That's how much respect Louie garnered, perhaps that's also why they had started calling me "HardWay Harris."

The call to 'bug-out' had come in the dead of the night. In the scramble, I somehow ended up in the driver's seat of a supporting APC, commandeered from the passenger side by Sgt. Louie. Louie was from another platoon, and we barely knew each other, but I definitely gave him something to remember me by.

I was slow getting the carrier started. I didn't drive it much and couldn't quite remember how, and Sgt. Louie was overly cautious about letting me merge herky-jerky into the convoy. Under the canopy of the trees, the absolute darkness was stifling. I couldn't see my hand in front of my face, and we wound up as the tail end of our stretched-out military parade. I pulled in right behind the very last vehicle and right off we were in danger of becoming the lone straggler.

I wasn't very good at driving that twelve-ton aluminum brick. The M-548s were a big and boxy variant of the classic APC. As far as track vehicles go, they are incredibly unresponsive with a large blunt wedge up front, just below the high-riding cab, and a wide flat behind. Also known as an all-purpose carrier, as adept with moving men as munitions, the APC was a cavernous metal garage on tracks with a small driver's compartment up front. And the size of it was daunting—I could have used more road time. But I wasn't allowed to drive much at all . . . and I'll bet Sgt. Louie could tell you why.

As part of the simulated war game, we were mandated to use our infrared headlights only, even in the pitch black of the Bavarian forest. Those ruddy lights barely illuminated two yards ahead, and I struggled to keep up with the water buffalo right in front of us. But, in a bump and a jump, the buffalo's twinkling taillight reflectors vanished. After ten minutes of seeing neither hide nor hair of our convoy, Sgt. Louie started to get antsy. We had no map or directions, and not even the faintest idea where we were headed.

We were completely on our own.

"Faster, Harris, faster," Louie kept insisting.

But I was pushing that big metal bucket as fast as I could, through zero visibility to boot, and neither one of us knew where the road went. I had my NCO repellent at the ready—an unlit Danneman cigar, which looked and tasted like rolled dog shit in my mouth, and the diesel pedal all the way to the floor. I was thinking *If only we had some tunes in this hunk o' junk* when the barrel of a Patton battle tank came out of nowhere. It sped straight at us like a giant skewer, threatening to make shish kebabs out of us.

"To the right, Harris! To the RIGHT," Louie bellowed in my ear, his voice full of panic.

The weedy backwoods road we were on had soft shoulders and steep drainage ditches on either side, surrounded by farmland. There was barely enough room for two normal vehicles to pass side by side, let alone a pair of Uncle Sam's overgrown ironclads. Again and again, out of the gloom came a fifty-two-ton tank with its barrel aimed right at us, barely missing us before disappearing into the shadows behind. One after another after another, a never-ending single file of Pattons flew by in the opposite direction. Someone had clearly called the cavalry.

"Over to the right, Harris!" Louie demanded.

From my perspective on the far left I could see that it was an optical illusion: the guns looked like they were coming right at us, when in fact they were passing to the side by a good six feet. Still, it was close enough for Sgt. Louie to keep yelling at me to get over. No matter that we were already as far over as possible without plummeting down the side.

"Harris, to the right!" Louie insisted again.

The tanks kept coming for what seemed like an hour. They weren't expecting to come across us, and most had to make adjustments at the last second to avoid clipping us. There was so little room for error that it was impossible to relax, even for an instant. I tried to keep my cool and maintain velocity, but my concentration kept getting shattered. My arms were screaming at me, my hands were exhausted, and my nerves were worn thin. And still Sgt. Louie kept shouting at me to get over to the right. Finally, I snapped.

"You want me to get over to the right? Then let's go to the right!"

I pulled sharply off the road and drove the APC head first into the ditch. With the tracks going at full speed, we had so much momentum that we dove down, hit bottom, then bounced back up. I could have sworn we caught air. When we landed in the field on the other side of the ditch's far embankment, I kept right on going.

Initially, Sgt. Louie was too freaked out to say anything, which was highly satisfying. But as we were plowing through German countryside, he started to scream hysterically.

"Harris!" he cried, "get back on the road! Harris!"

I was laughing madly, thinking *there's no pleasing some people*. It felt like I was caught in a grand theater of the absurd and was trying to break out. Surely, at some point, I'd burst through the backdrop of that topsy-turvy reality and be free.

It was a thrilling and terrifying ride at break-neck speed in that huge clumsy vehicle. Except for our infrareds, the world was devoid of all light. Fences, trees, small shacks, and whatever else lurked in the fields would appear in front of us for the briefest moment before we smashed through it on our wild way. Neither Mr. Toad nor Sgt. Hillbilly had anything on me that night! Sgt. Louie exhausted himself as we tore through the hinterland, going from screaming to pleading with me.

"Please, Harris, please . . . please pull back on the road."

The moon peeked out from behind the clouds, and I saw tears streaming down Louie's dusty face, leaving shiny snail trails on his cheeks.

Have you ever had a grown man sitting next to you blubbering and begging? It's not a pretty sight. I didn't enjoy tormenting Sgt. Louie; he was a decent enough guy, not much of a soldier but pretty harmless. I didn't mean to break him like that. He just really got on my nerves. Thankfully, he wasn't yelling at me to get over to the right anymore, which was a sign to me that he had probably had enough.

I veered swiftly into the embankment we were paralleling and hit the ditch at a hard angle to get back on the road. For a second it felt like we might tumble backwards as we rose up and out of it, but APCs are like "Weebles"—bottom heavy. Propelled out of the ditch as if in slow motion, we hit the road askew with a clamor and slid. Through the panicked flight I still had the tracks going full tilt, and the moment we regained our traction the tracks grabbed hold of the ground, jerking us forward once again. I pulled back hard on the handbrakes like I was reining in a leviathan and we came to an abrupt, shuddering halt. The darkness felt abysmal. I quickly flicked the regular headlights on and off to get a better sense of our bearings, and there, not more than five feet away, stood our commanding officer, Lt. Colonel Jerkes. His back had been turned toward us, but he spun around at the clang and the flash to find our heaving hulk of metal looming behind him.

His face registered more than surprise. I think he knew how close he came to being another casualty of the Cold War. If we had kept moving forward, if I hadn't hit the brakes at the precise moment I did, he would have been smooshed like a two-hundred-pound packet of ketchup. The boars were always around to pick up any leftovers, making me wonder if we would have found parts of him when the sun came up.

The colonel and I stared at each other intently, each appraising the situation. Finally, he stepped aside and I slowly pulled the vehicle into position. Sgt. Louie didn't say another word, and I never drove an APC again.

*March 14, 1985*

Hi Harvey,

Did you hear that Chernenko is finally dead? We're on alert out in the field waiting to see if something happens. This could be a game changer—that guy was a fucking dinosaur. I'm betting they go Crazy Ivan on us before it all goes kaput and they make a major land grab. First they'll suck back up Yugoslavia and Albania so they can go from the Black Sea across to the Mediterranean. Then they'll go after Finland. And if they take over during the winter, no one will notice until spring. We'll see . . .

Thanks for the letter about Esther. It's hard hearing about this kind of family news long distance. So she got married to a guy we've never met—and you were the only one out of the whole family to go to the wedding? What the fuck? And you "gave her away" . . . in a church? That must have felt so weird for the both of you. Did you guys have to kneel and kiss the priest's ring? Do they do that? Did you eat one of those wafers? You know Jews have been killed for less. And did you drink the blood (was it really Manischewitz)?

How was the reception? Did they have a whole roasted pig? 'Cause I know that's big with Filipino weddings and a whole roasted anything is so freaking awesome. The only thing better than one big pig would be if they were Cornish hen–size so everyone could have an entire little piglet. Wouldn't that be cool? Remember the pickled ones we had to dissect in biology? How about those? They would be the perfect size—I say poach them in hot sauce and butter and then finish them under the broiler. If I think about it hard enough, I can almost taste it.

Sorry, but I am so hungry. I keep a bottle of Tabasco in my first-aid pouch, and whenever my stomach starts to growl I just put *a couple of drops on my tongue. My mouth lights on fire, my belly clenches, and I forget about running on empty for a little while. Trust me—it works.*

*One of my buddies got an Article 15 for trying to shoot a boar using a blank charge and a cleaning rod. He almost got the sucker. Apparently the wild boars are riddled with parasites. Or maybe they tell us that so we all don't start eating them.*

*I got my first shower in a month last week, and as I was walking out of the cinderblock shower house, a dozen striped little piglets ran right past me not more than three feet away. I just about shit a brick. In all my life I've never been so scared of anything so cute. If their big momma had seen me, and you know she had to be close by, it would have gotten ugly fast. The shower is just a single square room with no windows and only the one entrance. If she had chased me in, I'd have nowhere to go but heaven.*

*Those adult boars are the masters of the Black Forest. They're about 500 pounds, as smart and hungry as we are, and don't back down. The adolescents are about half that size. We make a game of chasing them into the tree line. They only run so far and then they turn around and chase us back. It's fun but a little scary. Imagine what the Army would tell somebody's parents if they got killed by a wild pig: "Sorry, folks. Your son was eaten by bacon bigger than your boy."*

*And I'm back to my food obsession—okay . . . a little more Tabasco. Chow isn't for a couple more hours and I have to carefully ration the food I have with me. The field makes every small thing so crucial, and surviving in the outdoors is never by accident. Dying can be accidental—but living is purposeful. I look out for #1 first and foremost 'cause no one else will. And I never lose sight of when we're supposed to get the fuck out of here. Hell, that's what keeps me going . . . knowing that there's an end to this.*

*Hey, I was expecting some hits in that last letter. What the fuck? I'm going to assume you're working on it. The check I sent should have covered it. And all you've got to do is stick it between the pages in a book. How about sending it to me in Fear of Flying by Erica Jong? I've heard it's pretty erotic, and it'll make up for your tardiness. Hopefully by the time I get your next letter I'll be back in civilization—clean and drunk. Amen.*

*Alright, I love you . . . and send me that shit,*
*Adam.*

# NINETEEN

⟷

## Gedunkeroo

Y ou'd never guess what is hidden in deep dark recesses of the
Bavarian Black Forest—things bigger and badder than anything the Brothers Grimm ever dreamt of, stashed away by Uncle
Sam and guarded by American GIs. My unit had just finished six
weeks in the field, and the rest of my battalion was going back to
the rear. Washington, the one other two-year man from our battery,
and I had been pulled aside and put on the back of a five-ton truck
with ten other privates from different batteries, all two-year men as
well. It was an undisclosed duty, but when you're a lower enlisted
it's always undisclosed. "Grab your gear" was all we were told—
standard operating procedure for the subordinate—then driven
deep into the heart of darkness.

It turned out to be a reunion of sorts for us "Minute Men,"
as we two-year enlisted were called. There weren't a lot of us, and
I ran into a couple guys from basic training I hadn't seen since
our days of blood, sweat, and fears in Ft. Sill. With our short-timer
contracts we were easily labeled the most disposable, and the
Army was going to get their money's worth out of us. So while the
rest of our companies returned to civilization for a well-earned
rest, we were given another month of living like animals in the
field.

Interred in an ancient one-room German Army barrack in the middle of Butt-Fuck Bavaria, we were off the grid and below the radar. Our cabin came without running water or electricity, just a whole lot of field mice for company. They were cute, but those little buggers got into everything and I constantly found their droppings in my boots. A single-seater outhouse and a permanently parked water buffalo were our only amenities. They were also the only other structures on the edge of an immense ammo pad, one of many secreted in the darkest pockets of the Black Forest.

Our job for the next month was to guard a concrete expanse the size of ten football fields, loaded with every type of armament and weaponry a person can imagine—and then some. Stacks and stacks of wooden boxes filled with explosives—rockets, grenades, and missiles for every occasion—were piled taller than I was. In other words, row upon row, crate after crate, of everything the US would ever need not *if*, but *when*, World War III broke out.

The official strategy was that soldiers close enough to the Fulda Pass, like those of us stationed in Hanau, would block that spillway into Europe with our dead bodies, allowing time for invading US troops to reach the supplies we were now guarding. The brighter ones of us knew this, but what really spooked me wasn't the Soviets—it was the German homegrown terrorist groups like the Red Army Faction, or as the briefing sergeant called them, "the Badder-Meinhof Gang."

Months prior, news of the RAF joining forces with the Belgian "Fighting Communist Cells" and the French terrorist group "Direct Action" had been splashed across the front pages of the armed forces newspaper, the *Stars and Stripes*, following the assassination of a German aerospace executive. The groups had pledged to target the NATO "imperialist military machine" throughout Western Europe, and that put us right in their crosshairs.

Then there was the spreading violence from the Mideast. Already that year, the Abu Nidal were implicated in the attack on a bar popular with US servicemen in Athens; Hezbollah, who had previously targeted Marines in Beirut, bombed a Jewish film festival in Paris; and the Islamic Jihad had blown up a restaurant frequented by GIs near Madrid.

Bombings had stepped up since the new year, and NATO installations, like the one we were guarding, were at the top of everyone's hit list. Imagine the coup if a terrorist group could take over one of our ammo sites. They would have all the explosives they could carry and at least a dozen dead Americans to flaunt. Then there would be the political fiasco after they exposed the site by blowing up what they couldn't cart away.

It was the first time any of us were given live ammo in the field away from the firing range—two big banana clips each—and ordered to walk "locked and loaded with safeties off." This was highly unusual during a peacetime duty and the only time it ever happened to me. The Army was cautious about giving privates bullets without a whole lot of supervision, so believe me, they weren't doing it to impress us. That detail, sure as shit, felt like the real deal.

They chose a total of twelve of us sad sacks to patrol, in groups of two, the perimeter of that enormous ammo site—both the hilly side and the wet and marshy side. It took four hours for a pair to walk around the entire compound, and patrols were sent out every two hours. The hilly side had a narrow worn path just wide enough for one, so we had to walk single file through it. It was bordered by a ten-foot-high chain-link fence topped with razor wire, which ran through the forest as our one and only line of defense. If someone wanted in, a pair of wire cutters is all they would need for admittance.

The forest was so immeasurably dark that we could walk into a booby trap and never know until it was too late. The half of our trail that ran through a marsh was bisected by six small bridges. If there was a fence on that side, we didn't see it. Washington and I were given the twelve to four shift—that's noon to four in the afternoon and midnight to four in the morning—four hours on and eight hours off, day in and day out, for the next four weeks.

As Washington and I got acquainted with our route, we got wise to two things. One, the hilly side was a blind trap; the only way to gain an advantage was to sit still and listen instead of tramping through it. Two, the small footbridges throughout the marshy side were lined with wooden rails that were the perfect place for one of my favorite beings, the European Orb Spider, to weave their webs. They are a beautiful molted shamrock green and pumpkin-orange color, large and crab-like with mild dispositions, and only one out of five—mostly the females—will deliver a nasty bite. As if to taunt us, their webs were evenly spaced along every rail, right where we put our hands to keep from tumbling into the water.

Without anyone around for miles to tell me not to, I kicked down every bridge's guardrail the very first day. That way, none of us would get used to using them and no one would miss them. Orb spiders often re-spin their webs every evening, so the displaced arachnids scooted off unharmed and only slightly annoyed. Washington watched wordlessly as I did this, trusting in my preparations. I reasoned that we would come to know that trail and terrain intimately, and that would be our salvation if we were attacked. For any intruder not as well versed in the path through those woods, the derailed bridges would be a sure liability in the dark.

Every third day, one guy from each pair of the roving guards would be allowed to catch a ride into the nearest village, where we could do laundry and buy what Washington, who grew up near a Navy yard in Georgia, called "gedunks"—snack food you might find at a Southern movie theater concession stand. Otherwise, we subsided on cold MREs in our lonely cabin cloaked by the Black Forest. It was like being in solitary confinement with eleven other guys. Between the monotony of our downtime and the high tension of being ready to shoot anything that moved, the detail started to wear on us all. I imagine at some point the Army played with the numbers and realized that five weeks on this grind was just too long and people would start to snap—but four weeks was doable.

Necessity being the mother of invention, Washington and I knew we were going to need alcohol to survive, so we came up with the "Gedunkeroo Plan." The one of us who went into town would leave his gas mask and first-aid kit hidden in his sleeping bag, and while getting his laundry done would slip away to the little market for supplies. It's surprising how much beer can be stuffed inside a gas mask carrier, with room for just one more in the first-aid pouch. Our rucksacks were checked for contraband before getting back onto the truck, but that's as far as the overseeing sergeant went—and that's what we were counting on.

During the day, our four-hour commute was a stroll through the woods, but at night that all changed. Blinded by the absolute absence of illumination, the forest path was an ambush waiting to happen. So we would sit in the pitch black at the top of the hill instead, our M-16s in one hand, finger on the trigger, and a beer in the other. Knowing we were a good two hours ahead of the next set of sentries, we'd sit and drink for an hour, just long enough to rehumanize ourselves.

At some point, Washington decided to teach me how to sing

Elvis. If someone had been close by, they would have heard the unmistakable sound of beer cans being gently sipped and the soft crooning of "The King" into the abyss—perhaps "Hound Dog" in an unhurried Southern drawl, interrupted only by the sounds and shrieks of the Schwarzwald dwellers.

On our last night of that God-forsaken detail, we were celebrating with an extra beer just past one in the morning, and it made us both tipsy. We were almost out of that hellhole and were giddy with the anticipation of going home the next day. Well, not home, but back to our post in Hanau where at least we had the promise of hot showers, mediocre chow, and women with lowered expectations.

Suddenly, off to our left was an indescribably horrific sound. The bushes near us began to move. Whatever it was, it was coming toward us.

We jumped up, our beers forgotten and fingers already exerting pressure on the triggers of our semi-automatic rifles. There was only momentary silence before a mewing sound mixed with sick laughter and murderous cries came from the underbrush a few feet away. Washington and I stood shoulder to shoulder, each giving the other courage and waiting for the worst. Whatever this thing was that had crawled out of the bowels of Bavaria, however big or bad, it wasn't getting past us without a fight. All our training came down to that moment and, sufficiently inebriated, we were ready to lay down a hail of gunfire.

After standing terrified side by side for several minutes, ready to shoot anything that moved, we finally caught a glimpse of what was haunting us.

Rabbits. Two of them. Fucking or fighting, we weren't sure which, with the eerie overtures of alley cats magnified one hundred times in creepiness.

We had very nearly shot that place up over a couple of horny rodents.

I don't even want to think what that court martial would have been like, or how big of a crater we would have left behind. We glanced at each other wordless, picked up our beer cans, and double-timed it all the way back to our cold-water shack.

— — —

On the way back to Hanau, in the rear of an old beat-up five-ton, the twelve of us looked like we'd just come from a month of funerals. Dirty, tired, and just plain worn out, our mood was somber like the jet-black forest we were finally allowed to leave behind. Instead of being elated at getting out of there, there was a perceptible and prevalent sense of mourning, as if those onerous two and a half months in the field had extinguished the joy in each of us. It felt desperate and I thought somebody needed to do something about it—so I took a chance.

Growing up with a speech impediment means doing things like not answering the phone or singing out loud unless you're alone. But on that bleak evening, for the first time in my life, I sang for the ears of others. The captive audience I had in the back of the truck had nowhere to go and must have been too fatigued to complain as I started to croon the lullabies my mother sang to me as a child— low and slow like Washington had taught me.

"I see the moon and the moon sees me,

shining over yonder sea.

Oh let the light that shines on me,

shine on the one I love."

Maybe that wasn't the textbook version, but it was how I remembered it.

Darkness hung around us like a mantle, the rumble of the truck tires over gravely road the only sound besides my voice.

"You Are My Sunshine" followed. Just like my mother used to do, I sandwiched the first stanza between two iterations of the first chorus: "And I hung my head and cried."

We couldn't see each other in the blackness, but the sniffles told me there wasn't a dry eye in the back of that truck. Somehow, our emotions ran thickly back into us, and when I was done singing, the pall felt tangibly lifted and the night no longer a source of oppression.

Hey Harvey,

Just got off work and cracked a cold one. The "stationery" that
I'm writing this letter on is the Duty Officer's Log—it's my job to fill
it out every day. And every day I falsify two whole pages of the log.
Not great big lies mind you, just the ones I was told to tell . . . just
doing my job. I mean, I'm an E-2. What the hell am I doing with a
Duty Officer's Log to begin with?

After we were done last time in the field, it was my good
fortune to be selected for an extra detail and I ran into an old friend,
Jefferson, from Basic. We were in the same Ft. Sill training platoon
and it's funny—we never really liked each other. But it was good
seeing each other again. Jefferson was always bragging about his
"size thirteen . . . if you know what I mean," and he said it so much,
it's what I remembered most about him.

The last time I saw him was a year ago when we were both still
newbies. It was the first time either of us had gone to a brothel and
we happened to be looking for the same whorehouse. We had both
gone alone for the sake of anonymity, but neither one of us could
find the place, so we looked for and found it together.

In Germany most of the prostitutes are housewives or college
students trying to earn some extra cash. When I went on the "fun"
runs on the weekend with HQ, I'd see them in bikinis sitting in folding
chairs along this one stretch of the autobahn. I had never been with a
hooker before but I must admit I was plenty curious. But when I
walked into the whorehouse with Jefferson that day, it felt more than
a little weird. It was the thought of having sex with a stranger while
completely sober, without any of the drunken flirting and foreplay
that usually goes along with picking up a chick at the bar.

*But just as we got there, a petite brunette walked in and made immediate eye contact with me—maybe she noticed I was staring at her. She was wearing thigh-high fuck-me boots, fishnet stockings, a miniskirt, and a tight sweater. Yeowza! She came right over to me and I (physically) couldn't hide my interest. Jefferson was talking up a couple of girls at the end of the bar and we hadn't even ordered our beers yet. But when she told me to follow her to her room, I couldn't say no. I felt a little nervous but was way more horny . . . and it was a pleasure to follow her from behind.*

*She had long black hair in really tight curls, green eyes, blood-red lipstick, and what looked like a great set of tits and child-bearing hips. I started to undress her—because even if you're with a prostitute, foreplay is still important—but she refused to let me take her bra off. She said her breasts were for her husband only . . . but I could take off her panties. I figured with the proper motivation, maybe I could get the bra off too.*

*Now, there are two rules when you're with a prostitute: don't kiss her on the mouth and don't go down on her. Well, one out of two ain't bad, and besides—I'm just not cut out to be a selfish lover. And she had just walked in the door so I knew she was fresh. I took the lead and went straight for the honey-pot. I started tonguing her clit while fingering her G-spot and she was really getting into it. You know if you do a low-volume hum, the vibrations from your lips transfer right to hers? And I was really working it. Just as she started to orgasm I felt her stiffen up, and I looked up at her and said, "I want you to cum in my mouth." You should try that line, it freaks girls out.*

*Sex is mostly mental and I think I really blew this chick's mind with that one. She grabbed my head with both hands and just buried my face into her as she came like crazy. Things got very juicy—and you know there's no faking that. When she stopped quivering and loosened her grip, I crawled up and slipped right inside her. She didn't even notice when I took off her bra.*

*Afterwards she made me promise to come back and see her again, but at the time I couldn't see paying for something that should be free— not on a regular basis at least. Now I get it. After a six-week stint in the field, saving up your ducats and storing up testosterone—it's a no-brainer to spend fifty marks (about $20) for a piece of ass.*

*When I was done I found my buddy at the bar chatting up the madame, and ordered a whiskey as an antidote for anything I might have picked up. Jefferson was bragging that the girl he was with was so impressed by his size thirteen that she didn't make him pay. Seemed to me she should have charged him double if he was really that big.*

*Well, that was about eleven months ago and I hadn't seen Jefferson since. He got picked for the same shit detail guarding an ammo depot, and it was like a little reunion. No sooner had we stepped foot into the cabin that was supposed to be our billets (really just a rough-hewn shed with cots) than a field mouse shot out from underneath the nearest bunk and streaked by us, making it for the open door.*

*Faster than fast, Jefferson's size-thirteen boot came down on that poor little thing and smashed it so hard that it just atomized. I wondered if it wasn't the size but maybe the speed of his size thirteen that got him comped that day. Well, you never know . . . and I'll never find out!*

*OK, take care and don't take any guff or wooden quarters. I'll "talk" to you later,*

*Love, Adam.*

# TWENTY

༺·༻

## Angry Birds

The chaplain was the single greatest waste of space I came across in the service, and that's saying a lot. But then again, none of the clergy I ever met would have been mistaken for one of "The Immortal Four." First, but not worst, of the chaplain's offenses was that he was an officer, and to make him even more candy-ass, he didn't even carry a weapon. So why was he wearing a uniform and how was he supposed to earn his chow?

The guy was supposed to be our religious and spiritual counsel, but that was only as long as you ate of the body of Christ—which did me absolutely no good. It may sound like nit-picking, but I never saw a single latke, hamantashen, or matzo my entire time in service. I tried to speak to the chaplain about it once, but he wouldn't give me the time of day. Sure as shit he wasn't going to go out of his way for the only Jewish soldier on post—oh, hell no. But you can bet every one of the Christian holidays were observed, and always by serving some version of pork at the mess hall.

"There you go, Jew-boy, have a little ham with your Christmas dinner. What's wrong . . . not hungry?"

I didn't like recognizing officers. How do you respect a man who doesn't roll up his sleeves, pitch in, and do some real work? And I really hated saluting the chaplain in particular. How the hell was he superior to me? But I knew as a captain he could, and would, throw his commissioned evangelical weight around if I didn't.

One day, as I was approaching Captain Chaplain, inspiration hit and it occurred to me that the greeting I was forced to utter while saluting him could be anything if done in the proper cadence. So, as we passed each other, I waited until the last acceptable moment to salute him, and then without warning I shouted as loud as I could: "Kill Them All and Let God Sort 'em Out, Sir!"

That poor bastard almost leapt out of his uniform in fright. He didn't see that coming, and I just kept walking—resisting the urge to turn and smirk at him. His reaction was all the reward I needed, and my duty was delightfully fulfilled.

It became a brilliant inside joke, like a grown-up military version of "I'm Not Touching You." I would think up new banalities to shout during the many monotonous hours spent in the motorpool keeping the "eight-inch" ship-shape. My friends thought it was hilarious as well, and while they didn't actively join in, they would give me suggestions on what to yell. Soon, whenever the chaplain saw me coming he would head the other way—which was just what I wanted: not to have to salute his sorry ass. It was so much fun that I decided to share my game with any officer I came across. The lower the rank, the easier it was to get away with.

My problem with commissioned officers was that I kept looking past the clothes. Instead, I focused on the person giving me the orders. Saluting an officer was like giving respect before it's earned, a process that has always resulted in me playing the fool. When saluting, an inferior must salute a superior as the two approach each other, and this must be done before their shoulders pass. The courtesy is then returned by a condescending salute from the officer to the subordinate, leaving no confusion who was the top dog—even if the higher in rank was a complete pussy.

I would wait until our shoulders would be side by side, my face as close to his as possible. At that point the officer would be impatiently waiting for his gesture of due respect, because no matter how young or stupid, he was still entitled to be called "sir"—a term that certainly did not apply to me, the low-life enlisted offering the salute. At the very last second before the officer had a chance to reprimand me, I would turn and give the greeting at the top of my lungs at point-blank range.

"Hostess Twinkies and Death—Sir!" was my favorite, though "Beans and MotherFuckers—Sir!" was a close second.

To me, it helped balance the slanted social caste system I had to live with. By protocol, the officer I had just made jump would have to salute me back—even after I had just startled the shit out of him. One shook-up lieutenant, who didn't even look old enough to shave, turned to me red-faced and teary eyed afterward. "Why did you do that to me?" he asked sincerely. I almost felt guilty. Almost.

I did it because the officers acted like they were too good to talk to me, to sit and eat with me, or to ever treat me like an equal. That was called fraternizing—as if I was the enemy. To them I was an indentured servant at best, a beast of burden at worst, without the intellect or emotions that make a person a real human being.

Occasionally I had payroll duty, which meant accompanying the payroll officer after he picked up the payday cash to deliver it to the bank on post. I would walk behind the officer carrying the heavy briefcase full of cash, with my M-16 locked and loaded.

When he went into the officer's mess for lunch, as he often did, and left the case by his seat, I stood by the table on guard. He never asked me if I had already eaten; in fact, he didn't say anything to me at all. Such a question would never occur to him when I was just a thing that came along with the detail.

So maybe that's why I did it—to make them notice me.

One afternoon, as I was heading toward the motorpool, I saw the post's lieutenant colonel entering the command building with three "full-bird" colonels in tow. They were quite a little club of up-and-comers; the next rung for each of them on the promotional ladder was the coveted rank of general. Our Light Colonel Jerkes was giving a tour to these visiting dignitaries, trying to impress them, and it piqued my curiosity about how the other side lived.

As I continued around the building on my way to the parked howitzers in the motorpool, I heard the heavy back door of Command clang shut. I looked and from my vantage point could just barely see the rear of the building. Jerkes and his guests had stepped out back for an after-lunch smoke—and none of them were wearing their head gear.

I instantly saw a once-in-a-lifetime opportunity.

I raced around the command building the opposite way I had arrived. Coming up on the four colonels from behind, I paused and peered around the corner. Sure enough, there was our fearless leader, enjoying a cigarette with his elitist buddies. One was leaning against the wall, two were sitting on the back steps, and Jerkes was reclining against the handrail.

As per Army uniform regulation 670-1, all servicemen were required to wear their head gear when outside, regardless of rank. But there they were, four big brass monkeys, smoking and joking like a bunch of yardbirds, doing what no enlisted man would dare.

With nobody to outrank them, they had complete impunity—who was going to call them on it? Rank may have its privileges, but even the high and mighty can trip over something small and insignificant.

With their defenses completely down, I blindsided them with a classic military maneuver.

"GOOD AFTERNOON, GENTLEMEN," I bellowed in my "voice of command," striding out of the shadows not more than twenty feet from where they were.

As if on command, all four officers jumped up and assumed the "at attention" position reflexively, like I knew they would. And like good little soldiers they all saluted me back—a lowly private with a grin so wide that if I'd been wearing lipstick, it would have gotten in my ears.

You can't imagine the dirty looks I got from these guys who made a living telling guys like me what to do. I never saw a single officer get his hands or boots dirty; that was for the enlisted. So maybe my hands were a little dirtier than most. I was just sharing what it felt like to have to jump when some schmuck with stripes shouted "How high?"

What were they going to do? Give me some crap detail? That's what I got anyway, and technically I hadn't done anything punishable. All I was doing was playing their game—except by my rules. And to think that only one year prior I had been the Most Distinguished Graduate, whose "attention to personal appearance, military bearing, and military courtesy were exemplary."

As Mae West said, "I used to be Snow White . . . but then I drifted!"

*Dear Harvey,*

*How's life treating you? Do anything for shits and giggles lately?*

*You ever wonder how a whole herd of howitzers gets across a country? On top of railway cars! It's a little easier on the German streets and a little harder on us. Last week while we were rail-heading, this guy ahead of us, from another platoon, was tying down an eight-inch. He was wearing one of those old metal helmets and when he stood up to stretch, he made contact with a voltage line running right above him. I was about two hundred yards down track, with my company right behind theirs, and we all heard this big crack. It sounded like an M-80 going off underneath a can.*

*One of my buddies is a medic and I've always been a little jealous because, aside from learning a trade that he can take into the real world (he wants to be an EMT when he gets out), he also gets to carry a sidearm, which is usually only reserved for officers. He was one of the first ones to get to the guy and I'm not so jealous anymore. Seems that the electricity turned the poor fucker into a crispy critter. It blew holes through the palms of his hands and soles of his feet, not to mention it tore the under-seam of his scrotum. I didn't ask but he told me anyway. And it's like I've always said: it's not the worst-case scenario without genital mutilation. So I think this qualifies.*

*"FISHDO," DumbOx said. "Fuck it—shit happens . . . drive on!" So we just kept moving and didn't think on it too much—that's all you can do. Add it to the list.*

*Get this, I got kicked out of my drug and alcohol program—or what the instructor called the "Denial Group." She's a "recovering" alcoholic, and if you disagreed with anything she said then you were just in the grip of Mr. Booze. She kept talking in the same tight little circles, trying to convince us that we were all alcoholics. She never did take into account the crazy world we live in, so I called her on it. She said I was interfering with the recovery of others and gave me the boot right before we went to the field. I was left wondering if there was another Article 15 waiting for me when we got back. But when we returned they just let it slide, or maybe enough time passed for the paperwork to be lost in the shuffle.*

*Then came the real punishment—a Company picnic . . . participation required! We were all piled into an old school bus and ended up being driven to Mainz for a tour of the churches. Turns out the chaplain had coordinated the trip for our spiritual edification.*

*So me and a bunch of my friends said "Fuck this—let's drink!" After a couple shots of German whiskey, my buddies and I settled down for some serious drinking and girl-baiting outside a small cafe. And they say we are America's ambassadors—whoops! Right next to our rendezvous point was a wine shop. We drank two bottles of a light and lively red between the four of us before getting back on the bus. All in all, I enjoyed the tour thoroughly.*

*A lot of guys were buying stuff to take home as souvenirs, but I'll carry those mementos in my head and in my heart, and save my money for Mr. Booze.*

*It's been so fucking hot and dry this summer, and Grafenwhoer gets so dusty. I keep having this reoccurring fantasy about the water buffalo—how our potable water gets trucked in. It's a big steel tank on wheels with spigots on both sides, and a large hatch at the top for refilling. As advance party in the field, that water buffalo becomes my lifeline. All I have to eat is dehydrated food, so I carry at least four canteens stashed in my rucksack and two hanging off my web gear.*

*I have this daydream about being all by myself and coming across a broken-down water buffalo. The rear axle's busted and the driver's gone for help. I strip naked, carefully putting my clothes on the branches of a nearby tree, and climb up onto the buffalo. I open the hatch and from my weight on the tank the water sloshes below. I can smell the water—it's like the scent of fresh cucumber and cut grass. I slip into it, hanging at first from the hatch opening, and then I let go, dropping down inside. The water feels so good. The splashes echo and the sun streams in from above . . . glorious! When I hear a five-ton rumbling in the distance, I climb out refreshed, dress, and melt back into the surrounding forest just as the repair crew arrives, leaving the water a little saltier. It's my favorite dream. Once in a while I fantasize that the driver, this tall skinny chick who I swear always gives me the eye, has stayed behind. We get into the water together and cluck like bunnies.*

*But honestly—it's the water and the bathing that I find the most compelling. It's like I crave it. That's the way it is with the very first shower back from the field. After weeks of smelling yourself and everybody around you, it's the shower and clean clothes that are on everyone's mind—even more than cold beer and loose women. And it's remarkable how when you wash off the layers of dirt, it's like washing off layers of animal. Your humanity returns, and the spark of civility you thought had gone out rekindles.*

*Take care of yourself. I'll be home before you know it.*

*Love, Adam.*

# TWENTY-ONE

~ıc~

## Bruce, Spring Us

In the summer of '85, American patriotism in the form of a buff Bruce Springsteen spilled over from the real world. When his Born in the USA tour came to Frankfurt, every GI in Germany wanted in. But I was stuck on alert along with thirty-nine other Death Dealers, guarding a nuclear missile silo. Where exactly it was, or whether it was a Nike, Minuteman, or Peacekeeper, I'll never know . . . or say. To a lowly private it made no difference. Three minutes to be in the back of a five-ton juggernaut as it roared out of the kaserne's front gates was all we needed to know. The U.P.s would get the call at the same time we did and often barely had time to stop traffic out front for us.

After the Soviets invaded Afghanistan in 1980, Souther pacifist Jimmy Carter reignited the nuclear arms race of the fifties, determined to burn the communist candle at both ends. Ronald Reagan then took up that torch with zeal, and within two years America and the Soviet Union had the nuclear equivalent of twenty million tons of TNT pointed at each other—with Germany smack dab in the middle. It was MADness.

Much of the far-left rhetoric painted the US as the instigator of the Cold War. But Reagan didn't overthink it; he was simply committed to leaving the USSR "on the ash heap of history." To help achieve that, Uncle Sam had ready-launch nuclear missiles peppered throughout five European countries as part of their fully automated strike against a Soviet "Dead Hand" nuclear attack. Humans had been taken out of the equation on both sides—they were prone to overthinking and could not be relied on to push the buttons necessary to destroy the planet in a thermonuclear holocaust.

Under those storm clouds of a Mutually Assured Destruction, the possibility of a final world war felt very real and spawned a lot of anti-American sentiment throughout Europe, which was not without merit. More than once Reagan had refused to engage in disarmament talks with the Russians, sticking to the dogma, "We win, they lose." As a result, beyond the peaceful protests and songs about ninety-nine balloons, there were plenty of homegrown German militant groups with far more radical ideas—and American missile sites were on top of their hit lists. So we never knew if it was just a drill, or if was the real deal.

Being on alert was a duty that never ended—it simply rotated from company to company, throughout the year and around the clock. For three long weeks at a time, forty enlisted men lived in the free-standing one-room cabin on post built for that sole purpose. There were no windows, interior walls, or dividers, just fluorescent ceiling lights, a wall-mounted TV in the corner, and a telephone by the front door. The closest things to permanent furniture were the forty well-worn cots that dated back to WWII and the unmitigated stench of body odor.

Our daily routine was ordered and unwavering. Fully armed and always in the same single-file order, everybody went to chow together, ate together, and left together. Only one guy was permitted to go to the bathroom at a time. When we showered every third day, we took over the first-, second-, and third-floor bathrooms of the nearest billets simultaneously. Otherwise, we all stayed sequestered like a cruel science experiment in a single room while tensions mounted and smells festered.

During those stagnant hours, we could untie our boots, but not remove them. We could take off our outer shirt, but our under-shirt had to stay on. We could undo our belts, but our pants had to stay up. Beyond trying to sleep away the endless monotony, we had two options for activities: playing cards and watching pornos. How guys could watch dirty movies sitting with a whole bunch of other guys was beyond me, so I played cards or tried to sleep.

Nobody was able to completely relax. We were activated every couple of days just to keep us on our toes. The one hundred and eighty seconds given to be deployed wasn't a suggestion; there was no wiggle room, and no matter how long a soldier may have been in, when the order went out, the adrenalin surged.

"Balls to the wall!" Sergeant Rodriguez liked to yell when the call came in. Within forty-five seconds we'd all be on our feet and quickly shuffling out the door—helmets strapped on askew, web-gear hastily buckled, assault rifles in hand and banana clips fully loaded. Any final adjustments, like tying one's boot laces, would be done in the back of the speeding truck.

Being pent up for weeks forces a healthy male to stifle all sorts of natural urges, and that leads to its own madness. Testos-terone was nature's way of limiting how long we could go without killing each other. At the start of the detail, we all resented the abstention from alcohol, but after a couple weeks it was a necessity as cabin fever started to set in. Disciplined sobriety may have been the only thing that kept us from turning on each other. By the third week we were all getting on each other's nerves and every little irksome thing became magnified a hundredfold.

I'd find myself thinking things like, *If that motherfucker trims his toenails one more time without picking them up afterward, I'll slit his throat with his own pair of clippers.*

Once the duty was finally over and we were let out of that rat-box, we were gifted with three days off to get reacquainted with the outside world—which meant seventy-two hours to drink, fuck, and fight.

Despite the isolation, our whole world had been buzzing about one thing and one thing only: the upcoming Bruce Springsteen concert. But getting a ticket to Bruce's sold-out show was nearly impossible—unless someone had connections. In the Army, in every battalion, there's always "The Man," as in "The Man That Can." He was the guy who could get pretty much anything anyone wanted . . . for a price.

After I gave up selling rice to the Korean restaurants, I kept a small hand in the black market with little stuff like Marlboro cigarettes (which were rationed), Cuban cigars (which in reality were Dominican), and commonly lost and found items like sleeping bags and gas masks. I also knew a guy who knew a guy who could get everything from weaponry to vehicles. So using up a few favors and twisting a few arms, I was able to score five tickets to the show.

Pug, Roland, and Slim were psyched—but Big Ben was ecstatic. Ben was from Asbury Park and considered Bruce his homeboy. I probably should have told them I had terrible concert karma, but it slipped my mind. Maybe the big batch of hundred-and-sixty-proof rum and pineapple juice cocktails we mixed up for the show helped me forget.

In a place where you can buy schnapps at the bus stop and beer in KFC, alcoholic drinks are always welcome, and we decided to start the party on the train. Most of the people onboard were going to the show, and we had come prepared. We handed out high-octane libations in Dixie cups to our fellow travelers while reserving the German version of "Big Gulps" filled for ourselves.

By the time we got to Frankfurt, the whole train car was happily drunk and singing Bruce Springsteen songs, with Big Ben as bandleader. The four hours we had to wait once we got there really screwed us, though. It was enough time for us to make a lot more friends and get really shit-face drunk. When they finally opened the doors, we were slurring our words and staggering. But we made it through the ticket booths and into the jam-packed inner atrium where, like grains of sand through a peppermill, people slowly filed into the stadium.

The number-one rule in Germany for all us partygoers was "Don't fuck with the Polizei." So when the mounted German police used their horses to keep people moving in the right direction— meaning by pushing their way through the crowd, not caring in the least if the horses stepped on people's feet—it was just too damned bad if you got squashed by a hoof. Limp but don't cry about it because they'd arrest you just for complaining, especially if you were a GI. So when Slim got stepped on he decided to do more than bitch. He decided to take revenge.

Slim had been standing backed up to the huge fountain in the center of the atrium and had nowhere to go when a mounted Polizei came riding right past and the horse trod on him. The outer lip of the fountain's water basin he was against was waist high, and Slim thought if he shoved the horse against it, he could topple both the offending animal and rider into the drink. All he needed was a good running start—except Slim was too drunk to run a straight line and plowed face first into the horse's ass. The horse jumped and almost threw the rider. I rushed over to Slim in an attempt to restrain him, hoping the authorities wouldn't feel the need to do so themselves in retaliation. But that was taken as interference, and it's never a good idea to come between the Polizei and their prey.

We were too damn close to getting in to the concert venue and weren't going to go down easy, so when they grabbed both of us, we were drunk enough to forcibly resist. But just as the Polizei drew their batons, Pug, Roland, and Ben stepped in. They managed to keep Slim and me from being clobbered, but the lot of us got arrested and hauled away.

The German police used these nasty zip-strip handcuffs to restrain us, the disposable kind made of thin plastic that have to be cut off to be removed. If you struggled they cut deeper into your wrists, and the Polizei only scoffed at our silent pleas for compassion—a trait that is not in their DNA, and being reminded of it only pisses them off more. They split up the five of us and tossed us into separate paddy wagons, each filled with similarly shackled youths and directed to different correctional facilities. Frankfurt's a big place and they had a lot of holding cells to fill.

Pug, Roland, and Ben, having more temperate personalities, were released by the end of the day and were back in Hanau that evening. Me, I may have been a little more belligerent. I thought it couldn't get any worse than missing the concert and being locked up in a German jail—until the guards decided to prove me wrong and proceeded to hog-tie me. When I got out of it, they tied me up again. The second time I got loose, they left me alone in my cell—but they wouldn't let me go.

Slim got more than a little Deutschland hospitality in the form of a thorough strip and cavity search. Late into the first day, he came shambling into my cell with a faraway lobotomized stare, refusing to sit down. I think they took the last of his innocence, but he never spoke about it. Labeled as the aggressors, we had to wait to be released to a commissioned officer. But nobody from Alpha battery was in a hurry to make the half-hour trip to get us back, so Slim and I spent the rest of our leave in Frankfurt without the scenery ever changing.

*Hey Harvey,*

*I got your letter and I understand where you're coming from, but I wish you would trust me on this one. Not only do they not test us for LSD, but you know ever since I was busted I get checked whenever there's a "random" piss test. And I've never had a positive. As long as I stay within tolerable limits I'm okay. And I'm going to trip with friends when we're all off post—so you don't have to worry about me shooting somebody. Alright? So please, I've sent you the money . . . send me the shit.*

*And now for something completely different from your man stuck in the muck . . .*

*I had such an awesome sight last night. As advance party I was camped out all alone somewhere close to the Czech border, and a little past midnight I heard something above me, like something huge flying overhead. I looked out my tent expecting to see the alien horde's mothership and saw what seemed to be countless squadrons of helicopters flying past in tight groups of 15 to 30. And they just kept going and going—I'm guessing several hundred in total.*

*What a bizarre sight, like huge mechanical dragonflies swarming through the black velvet firmament. I watched, knowing this surreal life into which I've painted myself is just a microcosm in the continuum. As long as man rules this planet, armies will march. I wonder how many battles have been fought and how much blood has been spilled in the very spot where I've pitched my tent. A lot of Russian blood by all the weeds, I'm guessing.*

*I just finished "A Mote in God's Eye." It's an excellent read and a great warning about the dangers of technology outpacing the beings that create it. Being around massive machinery like our* howitzers, it's easy to feel irrelevant in comparison. The really scary thing is if just one gun bunny went overboard and began firing rounds across the border, WWIII would start before anyone could stop it. A domino effect with nowhere to run.

The next morning I went rabbit hunting. They are all over the place and look healthy as hell. I heard munching outside my tent bright and early and saw half a dozen plump jackrabbits eating breakfast right outside. I would've shot a couple and eaten well, but they don't give us ammo. I threw my canteens at them hoping to brain one, but they're just too damn fast.

Fresh meat would have done me good. Instead it was "the Four Dicks of Death." Gummy hamster meat hotdogs from my Meal Ready to Excrete . . . so so nasty. And they don't even give you any fucking mustard or ketchup to get that taste out of your mouth. My Basic training battalion got the very last of the c-rats and the first of the dehydrated meal prototypes while we were still in Ft. Sill, and it's obvious that these are still a work in progress. I'm betting some robot designed these MREs strictly for caloric intake without taking taste or texture into consideration.

Do you know what it's like to eat only processed foods (with no life left in it) and nothing fresh? You know what that does to your insides? It's like living off taco meat every meal for a month. I'd kill right now for a nice green salad or a piece of fruit. We ran out of water last week and I had to live off of "Parmalat"—an ultra-pasteurized shelf-stable milk product in little cartons. It's like drinking bland liquified cheese. I keep a shitload in my rucksack just in case, but it's my last go-to.

After five days without water, it started making me a little crazier than usual. I had an unrelenting thirst day and night—and I *just couldn't take it anymore. So when we pulled into position around three in the morning last night, after getting set up, I cornered Cookie in the back of his mess trailer. He's the sergeant that makes our chow, and he knows me 'cause I've bunked with the cooks before.*

*Cookie's a really good guy. He's a buck sergeant with twenty years in, busted all the way down for pissing in the coffee of an uppity captain a year ago. He handed me a big ol' can of pineapple juice—he must have seen the desperation in my eyes. I heated it up and sipped the whole thing like warm soup, and it was exactly what I needed. In the morning I took the best shit in weeks and felt so much better. It seems like when the necessities get scarce, so does empathy—and a little compassion can go a long way.*

*We've got some stuff to talk about, so on the chance that the world doesn't come to an end anytime soon, I'll give you a call in a couple weeks when I get back to the post. And you know what I want to discuss!*

*Love, Adam.*

# TWENTY-TWO

><~~

## Ripple Without Rhyme

Spanning June and July of 1985, the USSR held the largest air and naval military maneuvers mankind had ever seen. The Soviets were going through a regime change, and hard-liner General Ogarkov, fresh from Afghanistan, had taken control of the Western Theater's wartime command. That's how the USSR saw it—as wartime. And scarier still, Ogarkov was enthusiastic about his role. The enormous staged blitzkrieg designed to destroy NATO entirely was openly rehearsed by the Soviets and included a mock thermonuclear attack on America, with Washington, DC, targeted specifically.

It looked like the Russian Bear was doubling down in its bid for world dominance. The size and might displayed by the Soviet armed forces was stunning and sobering, a scope beyond what was thought possible—and that didn't take into account the half a million Soviet troops stationed in East Germany. The security of the free world was threatened like never before and NATO felt it had to respond. A meager one hundred thousand troops answered the call and mustered in West Germany, preparing for the worst. I was one of them.

Somewhere in the desolation between Germany and Czechoslovakia, Private Leroy Spooner and I were waiting for the guns to arrive. There was a late-morning drizzle; the day was suicide gray and would stay that way. Leroy and I were hunkered up on a boulder at the opening to a large muddy clearing pockmarked with puddles. Beyond the clearing, the forest fell away in gentle foggy steppes and gave up a beautiful panorama—a pastoral scene hidden in the low-lying clouds. Overcome with fatigue and hunger, neither one of us cared about the view, though. We kept our backs against the large rock as our wind-break, waiting for the artillery to show.

An M110 howitzer is a big machine and it takes a full crew to keep her running. An eight-inch can travel up to thirty-five miles an hour in optimum conditions, but through rough terrain it's slow going and maps don't always tell the truth. This is why it pays to have somebody go ahead on foot to lead the way as the "advance party."

During exercises, or in the event of war, every gun section would send out a lone soldier to slip through guarded territory unseen, to seek out and identify any forms of resistance days before the gun got there. Usually, the advance party was the one lower enlisted the gun sergeant disliked the most. So of course, that was my job.

Howitzers always traveled in a long linear herd with their APC counterparts and five-ton trucks in tow, and the rumble of those mechanized beasts could be felt miles away. With all that exhaust alone, they could be easily spotted, whereas a single soldier was nothing but a fleeting shadow leaving little to no trace. Being all by oneself meant the advance party had to be self-sufficient; like a snail with its shell, I carried everything I needed on my back at all times—maybe a little more than most. From my past experiences I knew if something went wrong, I was on my own, so I was prepared for every situation.

Crammed with every necessity I could fit, my rucksack dwarfed me. At the bottom sat a pair of "emergency only" boots that never, ever got worn, along with a MOPP suit in the event of nuclear fallout or biochemical warfare. Next were essential comforts that could always be used as currency in the field—long johns, socks, and underwear. In the best conditions the same clothes were worn for days, if not weeks, on end, making a clean pair of underwear a prized item. Then came loads of toilet paper, enough to last for months. Everybody has their limits, and I knew mine. I could live outside with all the hardships that went with it just fine—even subsisting off those repugnant field rations—as long as I didn't have to wipe my ass with leaves or snow. That's where I drew the line. Of all the gear I carried, the T.P. took top priority. Piled on that were a bunch of shit-brown MREs. And then the top was packed with my small tent, folding cot, bedroll, and sleeping bag, as well as my plastic poncho, which was also the waterproof outer shell of my tent when it rained. I had everything I needed to feel right at home—a traveling shell that was all that could be seen of me from a distance.

Crouched there with Leroy, water dripped off the brim of my helmet and rolled down my poncho; puddles formed at the base of the boulder. The morning's somber silence, and my thoughts of the beach, were constantly interrupted by Leroy's lamenting stomach, audible above the static of the rain. To stay warm I closed my eyes and imagined how the hot sun would feel against my face. I could smell salty air and hear gulls cry and children call for their mother's attention—all with the rhythmic waves pounding in the background. With each suggestive layer, I would feel my warmth grow. And then Leroy's belly, sounding angry as if it was eating itself, would pull me back to reality.

Leroy and I were friends. I knew him well enough to know

that he started calling himself "Le Roy," and I was concerned about him. His eyelids were droopy and his cheeks were sallow. He looked like he had lost about fifteen pounds in the past five weeks we had been in the field, which was more than usual. A person needed all their energy to stay healthy living like we were, and there was no getting around it—I knew he was hungry.

I had more than fifteen food pouches jam-smashed into my rucksack, but food was a tight commodity and even kindness can be a fatal flaw. If I fed the guy next to me, all he'd learn was how to be dependent. And if the balloon really did go up, he'd be in a world of hurt. So I just sat there and didn't say a thing. Besides, it's not like he was going to die of starvation. He could probably get some chow when the guns rolled up . . . and I had to stay ready for the unexpected.

We were all issued one Meal-Ready-to-Eat for lunch, and as the advance party I got two more for every day I'd be on my own. The best part of those MREs was the packaging: lightweight plastic, air- and waterproof, and nearly indestructible. They were as practical as the MREs themselves were unpalatable, which made conserving food easier by lessening the temptation to snack.

Every time we got back from the field, I stashed every uneaten item and unopened meal in my footlocker—which meant I always went on maneuvers with a bank of at least ten MREs. Even if I hated a particular food item, I knew at some point I'd be hungry enough to choke it down. And by eating everything I could, I was able to maintain weight while most guys lost ten pounds. I even consumed the revolting turkey à la king and the reviled slab of fruitcake . . . oh, the horrors. I defy anyone to correctly identify in the dark the packet of peanut butter versus the packet that claimed to be squeezable processed cheese.

During a mission I'd get my MREs and my coordinates, then

set out ahead of the rest of the camp just before dawn. Without the fanfare required of that traveling artillery circus I associated with, I'd slip into the early-morning mist and disappear. Using my compass and topographical map, I walked through, over, or around anything between me and my rendezvous point. If I ran into anything the gun couldn't get through, I'd find another way around, blazing the trail with red cloth markers—like the rags you might see at a mechanic's shop—from a wad stuffed into the side pocket of my rucksack. Macabre red is, quite naturally, the color of field artillery. Either way, my final coordinates, like the clearing that Leroy and I were sitting at the entrance of, stayed the same, even if the paths that led there deviated.

A person can easily walk three to four miles an hour, slower through rough terrain and slower still with eighty pounds on one's back. On my own schedule, with nothing else to do but tear down the tent at sunrise and set up camp an hour before sunset, hoofing twenty miles between the twilights was reasonable. Once a week I'd rendezvous with my company and get a hot meal if I was lucky—though I was rarely that lucky. I was also supposed to get a shower twice a month, but I was never lucky enough for that either.

In spite of it all, advance party was a cathartic pause from my usual micromanaged military day, and I couldn't have survived the Army without the peace of mind that came with it. I relished the solitude of the Bavarian Forest, not bothered in the least to be all by myself in the woods. It reminded me of the years when my brother and I spent entire summers in a small tent by the shore of one Ontario lake or another, with a rowboat as our only means of transport.

But the sparse woods of the Algonquins are a far cry from the towering Schwarzwald forest, where the trees canopy a hundred feet above the ground. Even during the day it's gloomy dark, and so pitch black at night that one's sense of sight becomes a liability, as your mind can play tricks on you. When you can't see your hand in front of your face, you start to doubt the reality of your hand altogether.

In my little tent, with just that single layer of canvas to create the illusion of shelter, I would end every evening with transcendental meditation. I had read about it in a *McCall's* magazine and my goal was astral travel—my only means of escape. Every morning when I woke, I had the time and serenity to analyze how I felt, carefully decide which dehydrated meal would be breakfast, and then be on my way. I always dreaded the end to this solace when I reached my coordinates—which was where I was hunkered down with Leroy, waiting for the rest of the crew to show up.

The guns had to be close. I couldn't see them but I could feel them. Another advance party emerged from the tree line, coming slowly into focus as he slogged through the muck to our perch. Then another arrived, followed by more cold, wet, tired, and despondent advance parties, until there was a small murder of us hanging about the large rock. No one spoke or called out. We all knew each other, and there was no need to expend the energy. We were going to need it soon.

It wasn't long until the guns came slowly rumbling and creaking in their standard formation. Each vehicle had a big number stenciled in orange on the front for easy identification. Mine was A-23, so when they got a quarter mile away, I jumped up, leaving my rucksack high and dry on the rock, and began waving my arms. The rest of the advance parties followed suit, except Leroy who sat unmoving until the last moment.

When the mechanized parade drew close enough, we all ran toward our respective howitzers, collected our directional sticks, and then ran ahead to pull them into position. At a loping trot I led the driver, Roland, into the clearing from where we would shoot. Once at our firing coordinates, I stuck the first pole into the ground to give Roland an idea where the rear of the tracks on his left would line up. The thick wooden poles had large metal rebar spikes coming out the bottom and were joined by three yards of rope. The second one got planted to line up with the first in the exact direction my compass showed me: our direction of fire.

Then I turned to face Roland, who drove the gun as well as launched the rounds, and guided the huge track vehicle into position. When you've got a thirty-ton weapon coming straight at you, you have to have more than good communication with the driver—you've got to have real trust between you.

As I stood my ground, bringing the howitzer into position, Roland slowed and purposely looked over to his right. I knew him well enough to read his eyes and followed his gaze. On the ridgeline above us was a jeep, a general's from the vanity stars painted on the sides. Roland looked me straight in the eyes and held up his hands, making a circle with his fingers and thumbs. I knew what he wanted me to do but it was risky as hell.

When you have a weapon that shoots massive bullets fourteen miles downfield, even the slightest deviation in angle can have catastrophic effects. Pulling the piece into place at "zero mils" meant the gun was perfectly lined up and ready to fire without further adjustment. That's what "on time and on target" means, and for artillerymen that's the holy grail. Using exaggerated gestures to make a real show of it, I directed the howitzer into position like an ant deciding where the refrigerator should go. At the last moment I had Roland stop short. I gave one quick adjustment then signaled with both arms bowed overhead, fingertips touching in a big circle.

"Zero mils!" I shouted loud enough to be heard over the thunder of the supercharged Diesel engines around us. No other gun section was even close to being in position. Roland and I stared at one another as if we were afraid to look anywhere else. It was a total bluff all to impress the big brass on the hill; we both knew we weren't in place. Not even close. If we'd gotten a fire mission at that moment, we'd have been up the creek, and whatever was down range would be blown to smithereens—and that was a lot to gamble on.

"Move out!" the sergeants hollered.

The general's jeep backed up, turned around, and sped off. We received our new coordinates, five miles away, and with a sigh of relief went about our normal busy day blowing up great big pieces of mother Earth.

As an ancient Chinese proverb says, "Under a good general there are no bad soldiers."

— — —

Shortly after we got back to the rear, a general's inspection was announced—that's the "GI" that puts fear into every GI Joe. In all my time served, basic training included, that inspection was the most nerve-wracking and anally retentive situation I ever experienced—in what was already an OCD clean universe.

Every square inch of our little piece of American Imperialism in Germany had to be spotless, inside and out. Not a single piece of paper or cigarette butt could litter the entire post. Doors and hallways got painted, and the floors were buffed to a mirror finish. Everything not nailed down was moved, cleaned, and put back. Common areas like the TV room and poolroom had to look like a diorama in the original packaging. Bathrooms, which were always kept clean with a rotating schedule, were sterilized to surgery room standards. Every wall poster had to be removed, except religious—as in Christian—ones. Our lockers and living spaces had to be organized in identical fashion, and every microscopic speck of dust and dirt had to be removed.

"All I want to see are asses and elbows," our top sergeant bellowed after announcing the inspection at morning formation, sounding like he was revealing a fetish.

The motorpool was no exception to this standard of spotlessness. The howitzers were scrubbed clean in the wash area and repainted, then the wash area itself was scrubbed clean. The oil stains on the motorpool floor, which was just a great big outdoor concrete pad for parking the howitzers, were picked up with kitty litter and a little dance. The pad was swept and mopped, and parking lot lines were painted, along with the number of each gun, all artificially nice and tidy.

In addition to this, every service person had to look sharp. Each of us was required to get a haircut, even though our hair was already too short. Battle dress uniforms had to be pressed and starched, and our boots spit-shined—and that was an order! But true to my rebellious self, I didn't feel inclined to dress to impress. My military career so far had been fucked and I was not in the mood to look pretty. Nor did I ever learn how to spit-shine my boots. The medium gloss I got from shoe polish buffed with an old cotton T-shirt was as fancy as I was willing to get. I sure as hell wasn't going to spend hours tracing tiny circles with a mixture of water and melted polish on the toe of my boot just for that patent leather effect—the boots were for combat after all.

When the big day finally arrived, there was still a ton to do. So, as the lowest-ranking private in my room, I was chosen to wait for the general's first walk-through. He would inspect the post first, billet by billet, then the motorpool and the howitzers, and then the grand finale—the standing parade of personnel.

The rest of the guys were down at the motorpool. Last-minute orders had them repainting the parking lines and our gun's number a different color for uniformity. So I was ordered to wait by myself in a room big enough to house four men, where the smallest bit of grit and grime can always be found if you know where to look. That was why nobody else wanted to be there. They didn't want to be held accountable and be the recipient of the shit-cream pie poised to be thrown in somebody's face.

As the general approached my room for inspection with his entourage of the post's commanding officer, my top sergeant, and a small host of lesser officers, I could feel the atmosphere intensify.

"Attention!" I yelled, standing rigid by the doorway as he walked into the room.

"Who are you calling attention to, Private?" asked the general, looking around.

I wasn't sure what to say, so I stated the obvious as we were told to—loudly, like I had a pair. "Sir, the room, Sir!"

Generals pull tremendous gravity, enough to unhinge commissioned and noncommissioned officers alike, and he definitely enjoyed that perk. But I couldn't believe it when that overstuffed turkey reached into his back pocket and pulled out a white glove. *Oh shit, here we go again*, I thought, remembering Mean Joe Johnson at headquarters. The general knelt down beside Kash's bed, reached underneath, felt around for a bit, and pulled out his gloved hand. The general held it close to his face, inspecting it like he needed glasses—and it was spotless. Kash was such a neat freak, and that day I loved him for it.

"Well done, Private," the general stated as fact, then strode out like it was his castle and the rest of us were his jokers. There was something familiar about him, though . . . as if I had seen him before.

After the billets and the motorpool came the troop's inspection. It was quite an awesome sight: an entire post of thousands of soldiers all standing in perfect formation. Wave upon wave, battery upon battery, platoon after platoon of soldiers, all dressed to the tee. Their uniforms were crisp and clean, their hats were blocked, and their boots gleamed like mirrors. It was perfection—but for one single glaring fly in the ointment: me. Sergeant Dumb-Ox couldn't believe his eyes.

"Harris," he barked, beads of angry perspiration welling up across his forehead, "you look like two hundred pounds of ass-holes smashed into a hundred-pound bag." But it was too late to do anything about it. The inspection was about to begin.

As the general paced his way through the ranks, knees locked, shoulders straightened, and chests swelled. It was like being inspected by a living military god who could see right through us. I was following him with my peripheral vision and was starting to regret my decision when he came down my row. Then the general stopped right in front of me, the only rumpled sad sack in wrinkled clothes out of the whole kaserne. He hadn't so much as paused while he walked past countless other soldiers, but I stood out like a sore thumb.

"Private," he yelled like a person yells at a dog that's not listening, even though he was standing right in front of me and I was at attention. "Private, can you explain your appearance?" he demanded.

I had had plenty of time to think of an answer to that while I wasn't spit-shining, but I never thought it would come from "The Man."

"General," I answered just as loud, trying to level the field and not sound so intimidated. "The dress greens are meant for show, to be polished and pressed. We are wearing BDUs and such things render this battle dress uniform less efficient. Always ready, General, and always prepared. Sir."

I waited for the shit to hit the fan. Nobody schools a general.

Then, out of left field he says, "Private, didn't I see you pull your piece into position at zero mils last time in the field? Wasn't that you, Private? Zero mils?"

"Yes?" I replied, almost questioning. "Sir . . . yes." I had no idea where this was going, but I was sure it was a setup for a serious smackdown.

"And Private, did your gun sergeant buy you a case of beer?"

That was a new one on me. I was still looking for the trap, but decided to play along. "Sir, no, sir."

The general stopped, and it was as if the rotation of the Earth suddenly ceased and a pall fell over everything. The birds didn't chirp and bees didn't buzz. The entire post was deadly quiet. Slowly he turned, leaned back, and looked down the long row.

"Gun Sergeant," he shouted at DumbOx.

"Sir, yes, sir!" came the distant sergeant's knee-jerk reply.

"After formation you will buy this man a case of beer," commanded the general without missing a beat. "A case, Sergeant. Do you understand?"

"Sir, yes, sir!" replied DumbOx, with decidedly less spirit.

"Good," the general said to himself and went on his way.

It was the damnedest thing. I thought I was going to get reamed in front of everybody, and instead I was getting rewarded with a case of cold ones? All because of what one big brass monkey thought he saw. Could it get any crazier?

— — —

Later that afternoon, setting the beer down on the locker at the foot of my bed, Sergeant DumbOx asked, "What did the general say to you?"

"He asked me if I knew who killed Tonto," I said, making it up on the spot. I couldn't tell him the truth about Roland and me faking the zero mils—I might not get my beer.

"Wha?" fell out of DumbOx's mouth, hanging open too far to finish the word.

"And," I continued, "if I knew it was the Lone Ranger after he found out Kemosabe really meant asshole."

"Harris?" said a confused DumbOx.

He didn't know what the hell I was talking about. But it did sound like some crazy Army shit a general might say.

"Never mind, Kemosabe," I replied under my breath. "Never mind."

*Yo Harvey Superfly,*

*Whats up? I'm kicking back with a little yac 'n coke—that's Hennessy cognac to you, homeboy. Very sophisticated. Are you hip to Morris Day? 'Cause if not you've got to check him out. He is a fucking riot! Around here if someone says "what time is it" and you look at your watch, you're considered uncool. The correct answer is: "Time to do the Bird," and if somebody disses you, just say, "Ain't nobody bad like me."*

*Good news—the Army no longer thinks I have a drinking problem. Now they're convinced it's all mental. I got remanded to a psycho-drama group to work on self-awareness. They're seeing if they can "correct" my thinking. It's all a bunch of liberal bleeding heart bullshit, but it keeps me off certain duties and gives me plenty of sham time . . . so I'm not bitching.*

*Of course it doesn't keep me out of the field. This last time, one of my odd jobs was to help a German field artillery unit pull an old broken-down APC into the fifty-cal target area. It seemed just like any other shit detail, but for some reason there was so much tension. All the German guys were dressed for combat: flack-jacks, helmets . . . the works. And grim as fuck. Like they were keeping something to themselves—'cause usually they're very talkative (they like to practice their English with us).*

*Afterwards I found out a bunch of their crew had got shot up by an overzealous sergeant at the fifty-cal range just the week before. The Sgt. said he had a jammed round and never heard the cease-fire call through his earplugs. He cleared the breach and just opened up on the guys down range—killing four of them. That's some serious perforating power. Those 50-cal slugs are the size of a big stogie. When I heard that I was like, "Fuck it, I'm not doing this again!"*

*But I don't have a choice—I do it or I get locked up. Where's the Denial Lady now?*

*I had a weird thing happen to me on the way back from Munich. It was late Sunday night and I had the railcar all to myself when this really attractive woman got on the train. She came over and, with all the empty seats around, sat right next to me. We made a little small talk, really just flirting with each other. I told her I was a GI (like she couldn't tell) and she told me she was a designer's assistant coming back from a trade show in Milan. She had a big bottle of Grand Marnier and asked me if I'd like some. I say "Sure," and she opens it up and just hands me the bottle.*

*Well, I'm not going to refuse her hospitality—so I took a sip and gave it back to her. She takes a sip and hands it back to me, and we proceeded to get comfortable. We were cozied up and I made a pass at her by putting my hand on her thigh. But she picks up my hand. At first I thought I had gone too far, then she started kissing my fingers. It may be the most erotic experience I've ever had with my clothes on. I was sitting there hard as hell while she totally made love to my hand. Then she whispers to me that she's on her way back to Prague and I should come with her. She makes good money and she's got a big apartment all to herself but she's lonely and needs a man.*

*Now I love it when a woman is straightforward like that and I told her that I'd like to . . . but I had to be back at the post Monday morning. I must admit I was nicely liquored up, horny as hell, and was really thinking about going AWOL. I figured I could take a week off work, bang this crazy broad, and then go back and get busted. It's not like they're going to promote me. I'm no lifer—and they won't kick me out over it. So what the fuck?*

*But something just wasn't right. We'd had a class on SAEDA (Subversion and Espionage Directed against the Army) and the whole thing just sounded too good to be true. A smart, well-off woman who can't find a man? As we were approaching Frankfurt, I got up to use the bathroom, and when we pulled into the station I got off the train without even saying goodbye. Hell, I forgot my jacket and left it there—just a cheap one I picked up on the economy anyway.*

*I keep wondering what would have happened if I had gone with her. Was she a Soviet spy? And what would they have done with me? My gut tells me I did the right thing but it's still fucking with my head. A weaker person would have done it for sure. I mean this chick was hot. I keep picturing myself chained up in a cold basement somewhere in East Berlin, never to be seen or heard from again.*

*But you don't have to worry about that. Dodged another bullet—and I'll write you again in a few weeks. Write me back, you bastard!*

*Love, Adam.*

# TWENTY-THREE

## Bundeswehr Games

The Death Dealer battalion didn't have a baseball or football team, but we did have a soccer team—sort of. It was formed by Major Brownose in an attempt to score points with his boss Jerkes, our bottle-cap colonel. Jerkes wanted the Death Dealers to be represented in the 16th Annual Sportsfest Competition hosted by our German Army "partnership" unit, a Bundeswehr Feldartillerie Battalion. And the enterprising major made it happen, with as few resources as possible.

We certainly didn't run battle-ready maneuvers side by side with our German Army counterparts. At best, they were only there in supporting roles—from a great distance—when we were in the field. After all, if the Cold War got really hot it wouldn't be their fight. That's why they were letting us Americans tear all over their country with our oversized vehicles of war. They had provided the pitch and it was now up to Uncle Sam to supply the bodies. But the German Army did excel at playing soccer.

Patched together like a quilt made by angry blind men not on speaking terms, our team was comprised of an odd lot of Canadian, Jamaican, Puerto Rican, Filipino, and German-American enlisted from various artillery companies within the François Kaserne. We didn't have a coach, uniforms, or even enough guys to field two sides for a proper scrimmage—just four orange cones and a ball. And therein lay the problem: soccer is a sport based firmly upon the synergy of the whole being greater than the sum of the parts. In other words, teamwork. So the fact that none of us knew each other and had no idea how to play together posed a hefty challenge. Hell, even I was just a frustrated rugby player who only joined to get out of morning calisthenics.

Like a jury where everybody speaks a different language, we all played varying styles without ever reaching a consensus—which meant that we were just a bunch of unorganized individuals running around without a single set play. But in short order we learned we were there to show, not to win. Major Brownose was also a politician.

To make sure we would never actually meld as a team, we met only once a week, on Monday mornings, which practically guaranteed we'd all be at our hungover worst. And this was only if we weren't in the field, which was half the time, or on some damned detail, which was most of the time. Hence, our mission wasn't so much to compete against the Germans as it was to lose to them badly. It made the occupation slightly more palatable for our hosts, and that made the brass happy.

The Bundeswehr Games also included a mini-decathlon as the pregame show, and we each had our own events. The team's goalie and I participated in the javelin throw. We would practice in the evenings in an old dirt lot by tossing the only javelin we had back and forth to each other. It was like playing mumblety-peg with a spear in diminishing light, and the better we got, the farther away we had to stand, as we could easily have impaled each other with a really good heave.

The games were held at the German Army Post in Koblenz,

and we American GIs were put up in an empty wing of their billets—the same ones that housed the sons of bitches who wreaked havoc on the world just a few decades ago, the artillery units that defended Hitler's machinations of murder. Here now were their garrisoned sons and grandsons, and while my culture begs not to punish a child for the parent's transgressions, I felt no brotherly love.

Maybe it was the chip on my shoulder, or maybe my slant on history. At the onset of WWII, when my uncle Stan and his three brothers left for England to join the RAF, their local town's newspaper headline screamed: "Four Katz After a Rat!" For me, it was always personal.

That Saturday things went just as they were meant to, starting with the track and field events. We tried hard, but they won . . . over and over again. Then came the main event, the soccer match. Like a top-notch college squad against a group of high school delinquents, they ran us ragged all over the field. They could have scored on us mercilessly, but instead chose to toy with us, only scoring when they wanted to take a break. We never even got a foot on the ball.

You know you're losing badly when there's no point in keeping count. But I refused to give up chasing the ball, even though I knew at the last second they'd just kick it away to a teammate. Perhaps it was more fun for them to humiliate us than take us seriously, or maybe they were on orders to keep the score down. Either way, it was clear that they were much better trained and that we were totally outclassed. But what really pissed me off is that they refused to do us the honor of taking us seriously.

That night, after closing down the local disco with the rest of the team, I made my way back to the Bundeswehr Army barracks. Everyone else had gone ahead and the place was eerily quiet.

Hung above their doorway was an old cannon ramrod, a little piece of their revered military history, a symbol of German manifest destiny. I paused before entering, as if held back by something. With no one else about, the words of the late great George W. Plunkitt of Tammany Hall fame rang in my ears: "I seen my opportunities and I took 'em." I snatched the ramrod, wondering gleefully why they hadn't bothered to better secure such a treasured artifact of war, and set about to see what mischief I could get in.

After I'd gone through several rooms, poking and prodding sleeping individuals without any decent reactions, I decided to call it a night. I was worn out anyway, so I carefully slipped the old ramrod into the bed of my favorite sergeant, DumbOx, who was there not to participate but to keep order. I think in his dreams he thought it was his M16. He cuddled up with it while I headed off to bed in my room down the hall.

The next morning I woke to a great hullabaloo. The German servicemen were all in an uproar. Someone had stolen their precious little stick! It was of course found within minutes, still in bed with the snoring DumbOx. When they roused him, he claimed innocence, but the Germans were as insulted as they were suspicious—and with Germans it's hard to tell when they're giving the cold shoulder or just being themselves. It didn't matter to me one way or another, but it did bother the ranking officer of our little expedition, Major Brownose. It also wasn't difficult to figure out who the culprit was.

Later that morning he pulled me aside for a little talk. Not knowing me very well, he started straight off with the "bad cop" routine, trying to scare a confession out of me.

"Do you know how deep a hole you've dug for yourself by desecrating our hosts' home?" he scorned.

I was not impressed in the least.

"Did they get it back?" I asked, blasé. "And was it broken?" He refused to answer me.

I hoped to arouse some sympathy, if not empathy. "How many Americans died at the other end of their guns?" He remained blank. "You're a college grad," I prodded, "you studied history." His face got redder. "Do you know how many of my family members died in this country?" My eyes welled up. "So you tell me, where's the injury? Sir."

As far as I was concerned, all those killed by the Third Reich were my family and there would be no apology forthcoming. I didn't think I had done anything so egregious. The major merely got up without saying anything, and I thought maybe I had reached him. But he couldn't prove it was me, and nobody was willing to rat on me. He had to admit there was no damage besides slightly bruised relations. Still, it was a flaw that took the shine off the major's otherwise perfect achievement. It irked him that I, a lowly private, stood up to him and was completely unrepentant for my crime—and that I was getting away with it.

The following week we got wind of an impending field equipment readiness inspection. And lo and behold, I was one of two lucky souls blessed with cleaning all the pots and pans used by the mobile mess kitchen. Cookware blackened for more than forty years was expected to look like new—but no amount of scrubbing with steel wool did anything to their solid carbon patinas. We tried everything, including attaching scrubbing pads to a sander and then a power drill, but to no avail; the metal just would not shine through. Unsatisfied, the major refused to release us.

It wasn't a matter of cleanliness. It was a never-ending, impossible task the major relished as he squeezed us under his big thumb. After the third weekend in a row of spending every bit of my free time scrubbing the same damn pots and pans without any significant progress, I snapped. If I had to play their game, at least I could make my own rules.

With the mantra of "Fuck the Major" running through my head, I went to a hardware store off-post and bought four cans of silver spray paint—enough for all the pots and pans, inside and out, so they would be uniform because that's how the Army likes it. I had learned my camouflage lessons well, and wouldn't you know . . . we finally passed with flying colors. Brownose was as pleased as he was clueless.

The next time we went to the field, a strange case of instantaneous stomach flu broke out. We had pulled into position and were standing in the snow, waiting for lunch to be served up, when Private Lemon (so named because he was permanently sour) started eating before the rest of us. I had just told Pug to stick with the MREs and not to eat the food from the mess truck, and both of us had our eyes on Lemon. We were less than twenty feet from him when he abruptly stopped eating and his eyes bulged suddenly in surprise. Two seconds later, food shot out of his nose and mouth while his pants filled with the last meal he had eaten—all with tremendous pressure.

Synchronized projectile vomiting and diarrhea were the hallmarks of whatever was "going around" and it hit nearly everybody, the major included. Fortunately, the malady didn't last long, and there were no serious repercussions. Everyone was too sick to notice that the affliction never touched Pug and me.

As soon as the pots and pans were washed well enough so that their bottoms were once again showing their dark luster, all signs of the illness ceased. Curiously, no change in the taste of the food was ever detected.

DEPARTMENT OF THE ARMY

HEADQUARTERS, 1ST BATTALION,

40TH FIELD ARTILLERY REGIMENT

3RD ARMORED DIVISION (Spearhead)

APO NEW YORK 09165-1558


14 August 1985


AETFOD-E-CDR

SUBJECT: Letter of Commendation

PV2 Adam Harris

A Battery

1st Battalion, 40th Field Artillery Regiment

APO New York 09165-155


1.   I am taking this opportunity to express my
     appreciation and commendation for your
     participation in and performance during the 16th
     Annual Sportsfest Competition hosted by our German
     partnership unit, FELDARTILLERIE BATTALION 545, 10
     August 1985. Your hard work, competitive spirit,
     and intense desire to excel made a significant
     contribution to our battalion's superlative
     performance. The DEATH DEALERS won this year's
     Sportsfest and your individual efforts had a great
     deal to do with our overall remarkable
     achievement.


2.   Again, I commend you for your diligence and
     dedication and I extend my sincere appreciation
     for your support and loyalty.

3.  A copy of this letter will be placed in your
    Official Military Personnel File (OMPF).

WILLIAM P. JERKES
LTC, FA
Commanding

*Guten Tag Harvey,*

*Check out this stationery. It's from Colonel Jerkes (rhymes with herpes). I was expecting an article 15 and instead got this letter of commendation! I was going to wipe my ass with it but it's not even good for that—too rough.*

*I've been playing soccer for the battalion. Our team is made up of guys from a half-dozen batteries. They tease me that I run like I've got a broken leg . . . I don't think my foot will ever be the same. But I can still run. And I'm starting—even though I'm not a soccer player. I guess aggression has its privileges. Really it's because these Americans play like pussies. I run around in my old rugby boots and give whoever has the ball one of two choices: kick it away or be kicked. Or get your foot stepped on—either way it's pretty effective. Hell, it's just fun to run around like a happy dog, and damn it, that's my ball.*

*We went to Koplantz last week for a tournament against a whole bunch of other German field artillery regiments. They're all nice guys (our job after all is to save their ass) and it was beer day and night. And good food too. The Bundeswehr soccer teams are in what's like a semi-pro bush league. They've got coaches and equipment, and they practice constantly and really know what they're doing. Which is being gracious as fuck and then beating the shit out of us. It's so infuriating.*

*Before the match there was a mini-decathlon, and I participated in the javelin throw. I can really heave that thing. I was aiming for the ump and almost got that penguin—but he stepped to the side. Still, it was a good showing and I came in second. Then came the soccer game and as always they made us look like fools, finally beating us 7-0.*

*To cap off the day, the Germans brought out a great big rope for a tug-of-war—which was a surprise and not something we were prepared for. We chose from our biggest guys but were still one short so I volunteered. As there was only one position open, I ended up at the very front of the line. We had size, muscle, and weight for sure but the Germans had synchronicity and training. They all pulled as one, and Baby . . . it's all about the timing.*

*There just happened to be a mud puddle between the two teams and we were fishtailing to avoid being pulled into it. I started trying to shout out a cadence to counteract the German team but the crowd was laughing so hard that they drowned me out. We put up a good fight of brute strength but their superior technique wore us down. And then as if on cue we were all simultaneously yanked off our feet and dragged through the mud. That's called a setup . . . or in German—schadenfreude.*

*Oh, the crowd roared, and covered in shit we gave our hosts a great big laugh on us. I looked up and saw our officers laughing at us too. They think they're so fucking superior . . . sipping their beers in the shade and making small talk like petty politicians. Do they think we're just dull clods? Clowns for their amusement? 'Cause that's how they treat us. Fuck every one of them!*

*We're heading back to the field in a few weeks and I'm still waiting on your letter. What was it Andy Warhol said about tripping? Every time you destroy your old ego and create a new one? I seriously think I need a new one, but I need your help. Say "Hi" to the girls in karate class for me and give Jerome a groin grab in my honor.*

*Take care, I'll be in touch,*

*Love, Adam.*

## TWENTY–FOUR

## Turkish Delight

Within an hour of Private King's arrival at the Alpha company's barracks for his cherry deployment straight out of Basic, he earned more than just our dubious respect. King was rewarded with a brand new nickname by Pug and me, the one that he would carry for years to come.

As part of communications, King was assigned to Pug's room, and before he had time to unpack, we sent him on his first detail—given by his ranking roommate, Pug, and instigated by me.

"Hey, Newbie, howja like to have your first German beer?" I offered. "I'm buying."

"Sure," said King, "that sounds great."

"Cool, but we're fresh out," I said, "so I'll give you the ducats to run across the street and pick up a case. Okay?"

"Hefeweizen," Pug chimed in. "Make sure you get Hefeweizen. And as your commanding specialist, you can consider this your first official mission as a Death Dealer." This didn't make any sense, but it sounded good.

The new private made it to the bar just fine, and from Pug's window, we watched him cross back over the strasse with the beer without incident. But it was on the stairs back up to our barrack's room that King slipped and earned his name. Rather than let go of the beer, grab onto the railing, and catch himself, he chose to bravely hold on to the beer and take a header right into the metal-trimmed edge of a concrete step. Technically, his plan was sound and he was successful in breaking his fall using his forehead—but what a mess it made.

King arrived at our door with an intact case of shaken beer and a huge gash in his head that looked big enough to insert a silver dollar into. A sanguine spurt shooting out with every beat of his heart had cloaked his face in blood. The CQ had heard him hit the stairs, followed the trail of red splotches, and came right up behind him as he stood in the doorway.

I quickly relieved King of the beer as Pug sat him down and the CQ called for the medic. Anyone who splits his head open in the course of getting you a beer and completes the mission de-serves compensation, and a cool nickname was the best we had to offer. To make it up to King, and prevent him from being labeled right off the bat with something derogatory like "Fubar" or "Bumble-Fuck," we baptized him "Crazy King." Then they took him to the hospital to get sewn up.

Crazy and I hung out together once in a while. He was cool, but it wasn't a perfect fit. I was the tail end of the Baby Boomers and he was the vanguard for Gen X. I didn't even recognize the names of the bands he listened to: Ratt, Poison, Anthrax. To me, it sounded more like pest control than rock 'n' roll. And he read *Cracked* magazine, which was a clearly inferior knockoff of *Mad* magazine. But Crazy had started dating my girlfriend's best friend, Betty, so we double-dated once in a while. "Crazy" may have just been a nickname for King, but his girlfriend really was more than a couple nuts shy of a pesto.

Betty was about ten years older than Crazy, but dressed like she was a teenybopper. She wore her face caked with white makeup powder and startling red lipstick, framed by bleached-blond Marilyn Monroe hair. I didn't know her history, and I couldn't see beyond her dark eyes, or read the self-inflicted India-ink tattoos of names written and scratched out on the inside of both forearms, but I could tell she was clearly haunted by something.

One Sunday afternoon, the four of us were at a little trattoria for lunch when a coke bottle fell off one of the tables behind us. Nothing remarkable, just a broken glass bottle easily cleaned up by the waitress. Betty excused herself and went to the bathroom, and I immediately got a vague sense that something wasn't right. The tone in her voice as she left, the way she moved—I couldn't put my finger on it, but something was off. My intuition was buzzing madly, so I asked my girlfriend, Claudine, to check in on her.

As soon as Claudine opened the bathroom door she let out a shrill scream. Crazy and I jumped up to find Claudine frozen in the doorway, pointing. In front of the mirror was Betty. She held a small shard of broken glass and her hands were smeared with blood up to the elbows. Something about the shattered glass had sent her over the edge—and I had felt it, without understanding what or why.

Claudine was wide-eyed, hyperventilating and nearly hysterical, but Betty just stood there with a faraway look in her eyes. Her muted lips were barely parted and her palms were facing us, displaying crooked red gashes that ran along the underside of each forearm, from wrist to elbow. Blood obscured the old tattoos and dripped bright crimson from her fingertips onto the white floor tiles below. The drops were so big they flowed into the grout lines and moved like eerie red streams. That's the image that stuck in my head. I instantly knew it wasn't Betty's first time cutting herself. And judging by the amount of blood, she was getting better at it.

— — —

Whether men come to accept that every woman is crazy, or women come to accept that men are all knuckle-dragging cavemen, it seems to come down to what level of craziness a person can put up with.

Claudine wasn't psycho like Betty, but she had plenty of skeletons in her own closet—starting with having four kids by four different men and not a father amongst them. Her first child had been taken away from her, and she was holding on tightly to the youngest three for the welfare check that came every month. She smoked, she drank, and she lived off government handouts. I was an uneducated, broke-ass private without a whole lot on the horizon, and it was doubtful I was going to stay in Germany. I was at the beck and call of the Army first and foremost, and frequently in trouble. It was far from an ideal situation for either of us. But having someone to hold and care for, to walk with and to talk with, is wonderfully humanizing. And that's what we both needed—desperately.

— — —

A few months after Betty cut herself in the restaurant, Crazy fractured his tibia. Rumors that he would neither confirm nor deny were that when he broke up with Betty, she went ape-shit on him and shoved him out of a moving taxi, breaking his leg. The "walking" cast he was eventually fitted with went from his foot all the way to the hip, and it bothered the hell out of him. When

Crazy was finally able to shed his crutches, he decided to go out to the disco with Slim and me. He picked an unusually warm and humid evening, and all the cabs were taken but we were determined, so we ended up walking.

By the time we got there Crazy was sweating up a storm, and after a few drinks he couldn't take it any more. That cast had been itching and abrading him since the day they'd put it on.

Crazy asked me to follow him into the bathroom, then he put his leg up across two of the sinks and turned on the water. As the plaster of the cast softened, he started to tear chunks of it off into the sink. I leaned in to help but the odor was overwhelming and I had to step back. Just then a bunch of drunken Marines burst into the bathroom.

"Hey, this guy needs help taking off his cast," I announced like a circus barker.

Marines live for a challenge, so my statement was like throwing a beef carcass into a pool of starving piranhas. Before I even left the bathroom, random guys started ripping hunks of Crazy's cast off his leg. The old expression "Be careful what you ask for" came to mind, because suddenly the choice was no longer his.

Crazy rejoined Slim and me at the bar a little while later. Though he had cleaned up as best he could, his appearance was nothing shy of bizarre. He had the shocked expression of someone having just seen their parents naked in bed; his hair was tousled and teased as if it had been yanked on from all angles; and the large pink scar running across his forehead was pulsating, standing out in contrast to his pale, sweaty forehead. His white button-down shirt was soaked and clung to his chest, revealing freakishly hairy nipples. And his pants, which had the right leg cut off just below his groin to accommodate the cast, now exposed his atrophied leg, which was glow-in-the-dark white and locked into a slightly bent position. To top it off, a pungent odor of funk, like bad fish sauce and baby vomit, emanated from Crazy's leg. All of this embellished the fact that he couldn't straighten or bend that leg, which resulted in a limp that should have been registered at the Ministry of Silly Walks.

When the club closed down, the three of us weren't ready to call it a night. For some reason, all the alcohol we had drunk still wasn't enough to attract any women—and we all had the same thing in mind. Fortunately, across the bridge from the disco, on the other side of the autobahn, was a Turkish whorehouse. None of us had been in that particular brothel before and we decided to try our luck.

I got to the door first and it opened, which is always a good sign. Toward the back, next to the bar was the madame. She was counting money, which is not a good sign. The place was dark and I didn't see any women hanging around.

The Madame looked up at us and pulled a newspaper over the cash. "No, no . . . closed," she said firmly. "You go now. You go now—we are closed."

Sometimes in life you have to ask twice for what you want, so I did. It was then that she stepped out from behind the bar. A German rottweiler the size of a Shetland pony stood beside her—a growling beast of at least a hundred and twenty pounds, on a very short leash. She started coming toward me, with the dog leading as if to scare me off.

Undeterred, I began once more to ask nicely if any of the ladies were still working. But before I could get the words out, she barked, "*Rouse, dummkopf, rouse!*"

Now, I don't like being called names or given orders, and I really hate being cut off—especially when I'm drunk. And attempts of intimidation tend to backfire when used against me. I didn't like the Madame's menacing attitude or the way her guard dog was showing me its teeth, and I sure as hell didn't like being shoved around. Mischief whispered in my left ear and I figured if I couldn't get laid, maybe I could have some fun.

Then the rottweiler lunged for my crotch. I know and love dogs well, but I wasn't about to be bitten in the balls. Faster than fast, I smacked the top of its big boxy head as hard as I could with my open hand. It made a loud clapping sound, like a balloon popping, and froze the dog in its tracks. Crazy and Slim were still behind me and in one fluid movement, I spun around and split the oblivious pair, running straight out the front door. I slammed it shut behind me and held on to it for about three long seconds. The dog was barking loud enough for the whole place to reverberate, and Crazy and Slim were pounding on the inside of the door for dear life. I yanked it open and Crazy shot out like a bat from hell and kept going without looking back. Slim was right behind him. I slammed the door again and held it closed till Slim got enough distance away and the dog had quieted down. Then I let go and ran after them.

When I caught up to Slim, he tackled me right off the sidewalk and onto the grass. We rolled down the embankment and had a good-natured wrestle. We were both drunk and laughing hard. One of the reasons Slim and I were such good friends is that he knew I was only joking and appreciated my slapstick humor. But Crazy—he silly-speed-walked his freaked-out ass all the way back to the post and didn't utter another word to me for the rest of the weekend.

*Dear Horse Race Harvey,*

*The end of the season is coming up and the Jays could go all the way! I'm sure if they get in you'll be laying down some serious dough? Freakin A . . . imagine a Canadian team winning the World Series! It's probably easier when you're not emotionally invested. Get this—I saw a German guy wearing a Chicago Giants jersey. Yeah, the Chicago Giants? Over here they have no clue. Then again, Europeans don't even know the difference between soccer and football, so what the fuck?!*

*I really miss things like going to a ball game. I love how America's might is shown through the pageantry of a championship game. The feeling you get when fighter jets do a fly-by—it's like Uncle Sam flexing his guns at the global beach. I miss sitting on butt-numbing benches, drinking tepid watery beer out of oversized wax paper cups and eating humid peanuts without a fucking care in the world. Oh, that's the life! I always thought the last two words to the national anthem should be "Play Ball."*

*But if I could change the anthem it would be with the addition of five words. Right after the part at the end, "Oh say does that Star-Spangled Banner yet wave," should be the line "You're damn right it does," spoken not sung. And then the part about the Land of the Free and the Home of the Brave. You know it pulls a tear out of me just thinking about it. Maybe I'm just homesick.*

*Well, I got a new job. It's an unspoken detail . . . the Army is full of them. Mac got transferred last month and I guess this is one of his leftovers. It's one of those unwritten things, but this particular duty can only be performed by an old private, and at 22 years, I more than qualify. Before a New-Bee goes to the field for the first*

*time, someone has to give him the "Don't Be Stupid, Stupid" speech.
I recall all the weird and freakish ways guys have been injured or
killed since I've been here. Nothing made-up or exaggerated . . .
that's not necessary. Just the bottom line. Someone always dies
when we go to the field . . . make sure it isn't you. We're here to kill
for our country, not die for it—don't be stupid, Stupid.*

*We go to the field for 4 to 6 weeks half a dozen times per year
and at least one person dies in every one of those maneuvers. But
you never hear about it in the news, do you? I think my boss, Ronald
(and I can't say a bad thing about him cause it's against the UCMJ)
is a master of mass media manipulation (say that five times fast). I
mean, that's his forte and how he got elected. Fuck, I think the
chimp could do a better job. Can you dig it? I know you can!*

*Hey, have you seen the movie Gremlins? I saw it last weekend.
Me and a buddy downed a six each, smoked a gram on the way, and
then laughed our asses off. Too funny, and those gremlins make
sounds just like what I hear outside my tent at night in the field.
Now I know what they are!*

*They call the movie "Eine Kleine Monsters," The Little Monsters,
over here. And Hanau is the home to the Bros. Grimm where they
still believe in witches, ghosts, and goblins—so you can imagine
what a hit the movie was. I'm pretty sure the whole audience was
drunk and it was more fun than I've had in a long time.*

*Hey, I'm half-baked and am going to keep on keepin' on until I
can't see straight . . . and the words are already blurring, so I'll
catch you on the flip side.*

*Love, Adam.*

# TWENTY-FIVE

## The Creator's Cloth

Like a blind man touching the tail of an elephant, the real danger in describing the Creator is assuming to know more about the whole than the little bit we come across. So without the blasphemy of presuming to know what God is, I will instead tell what I firmly believe it isn't. It isn't human, it doesn't have a gender, and it doesn't have petty human emotions like ego. It doesn't care by what name it's called, in what language it's prayed to, or if we believe in it at all. In fact, human concepts of right and wrong, fairness and injustice, do not apply to the Eternal. It does have an agenda but, as in the military, the mission is undetermined and the destination unknown. In other words, explanations are rarely, if ever, offered and we are too small to see the big picture.

We've been told that dinosaurs roamed the world for over three hundred million years, that huge water scorpions were the masters of the seas for some two hundred million years, and that giant terror-birds ruled the Earth for sixty million years. In contrast, Homo sapiens have not even hit the three-million-year mark. So the thought of us being tolerated by this planet for another fifty-seven million years, just to tie the birds for third place, is ludicrous. To me, it seems presumptuous to say that we are, or have ever been, the Eternal's favorite children. As biologist J.B.S. Haldane observed: "The Creator must have an inordinate fondness for beetles. He made so many of them." Human beings are not nearly as elaborate—and there are not nearly as many.

— — —

I've often wondered why I saw what I saw and why it was revealed to me. Was I supposed to tell people about it, or keep it to myself as I have done? Was it simply a test, an initiation, and was the ensuing enlightenment my reward for passing? Or will it be necessary for a future event?

It used to be if a person had a spiritual vision it was interpreted as a sign of divinity. Often the affected individual would be revered, as if imbued with whatever holiness had passed through them. Nowadays, people who have these experiences are frequently labeled as having mental instability, or believed to have had a psychotic episode requiring hospitalization—the kind that specializes in pills and padded rooms. To be clear, I'm not talking about the type of "vision" that follows a good blow to the head, but one that happens during full sobriety in the middle of a normal day. Well, as normal as it gets when you're in a hibernal hinterland, playing war games with tens of thousands of other guys along the Soviet border where Czechoslovakia's nose pokes into Germany's fat ass.

Along with the rest of the Death Dealers, I was deep in the field where internal clocks are kept by the rise of the sun, the smell of the mess wagon, and the grumble of your belly. In the field, nobody ever voluntarily missed a meal, unless they were too sick to get up. Our schedules revolved around chow-time. Breakfast was finished being served just as the sun came up, lunch always seemed to come too late at noon, and dinner was ready as the sun was calling it a day. Every single meal featured beef of dubious origin, stale bread, and frozen or dehydrated potatoes in one form or other. Creamed chipped beef on toast, otherwise known as shit-on-a-shingle, was the best thing on the menu, so it's no wonder we called the food "grub."

After pulling into position and getting set up, we had about an hour before evening chow, and I lured Pug out into the frigid elements with the promise of a little hash. As we gained distance from the camp, we quickly found ourselves stumbling into a complete whiteout. We had had a cold snap and it had snowed unusually early—but this was something completely different. It was like being in a huge snow globe, with the wind blowing in all directions and the snow swirling madly about us. Walking blind with our eyes wide open, there didn't seem to be an up or down, or a side to a side. As any sane person would, Pug stopped in his tracks.

"Don't worry," I reassured him, "I'm Canadian," as if that gave me some sort of advanced wintery perception.

We were making our way into the pitch-white solitude, completely sober with my hash still tucked away, when all sense of time and space suddenly slipped away. I had the distinct impression that we were being swallowed up as we walked for what must have been at least three quarters of a mile, feeling as if we were being drawn toward something specific. A much denser pocket of fog stood out distinctly from the flurry all around us, and as we approached, it suddenly got much colder, like we had stepped into the epicenter of a vast heat drain. And we could both feel something else too.

As real as the weather, a tremendous sense of tangible dread enveloped us, yet we continued moving forward. Within seconds, we discerned a small shrouded edifice. It was a deer stand, a wooden structure built for the sole purpose of easily slaughtering the unsuspecting.

The concentrated essence of pure evil became intensely perceptible. We both felt it caressing us, sizing us up, as if it had been waiting for us. The sensation filled my entire mind and body. I've never felt anything like it in all my life, not before or since, and I realized I was grinding my teeth, trying to chew down the terror. I turned to Pug for reassurance and found him staring at me for the same empathy. The look in his eyes told me he felt exactly what I did. Whatever it was, it possessed an unrelenting gravity, like a weight both our lives were being measured against.

We were pulled toward the trap door, and I knew in my core that something sinister was up there—something beyond explanation. I didn't want to go near it, but I felt compelled. So I reached out and put my hand on a rung of the ladder, as if that's what I was supposed to do.

Wordlessly, Pug put his hand on my shoulder and pulled me back. The look of fear in his eyes bordered on desperate. *How could we both feel the same thing, at the same time, without a single utterance between us?* I wondered. I took my hand off the ladder and we cautiously backed away. Then, true to my pledge, I led us safely out of the separate reality of the whiteout.

Just as we got to the edge of the fluid ivory cloud, where we could see our battery's encampment ahead, I stopped, my promise of getting stoned completely forgotten.

"I've got to go take a shit," I said. "You go ahead and I'll catch up."

As Pug disappeared, I turned around. My single greatest fear had been challenged, and I needed to know why. Perhaps it touched a nerve, something from my childhood that I thought I had put to rest. What I was most afraid of was cowardliness, of being so scared that I couldn't, wouldn't, fight back. To be so consumed with terror that I would abandon my self-integrity. I had never felt absolute fear like I had at the deer stand and I had to know which was stronger—faith in myself or fear of the unknown.

I plodded through the blinding snowstorm with a lump of dread in my throat until I reached that same ultra-dense pocket of fog. My heart was beating so hard that it was filling my ears. I put my hand over it as the atmosphere of horror and revulsion intensified and I forced myself forward.

For a second time I walked into the heart of that climatic being—only nothing was there. The stand was gone. Just then, like a soap bubble bursting, the thick fog pocket dissolved in front of me and in its place was a vision of the Earth, as if from very far away. I could discern the continents, the oceans, the mountains, and the clouds high above them. I squinted for a closer look and saw brightly hued bands of light, some more muted and others brilliant like Day-Glo, swirling about the planet. Those vibrantly colored threads interwove a spherical tapestry around the planet, cocooning it. Suddenly, I understood what I was seeing. The ribbons of illumination were comprised of the life energy from all the living things that were or would ever be. Like all other energies, such a thing is neither created nor destroyed, only recycled—to return to the Eternal as one synergistic entity.

It made me think of how a sea sponge operates, as if it were a complex organism, though each and every cell is actually its own unicellular creature. Aggregated together, each individual surrenders its identity to form a much larger multicellular animal with synchronicity of the whole. This allows the sponge to act as a single being with one unified purpose. Before my eyes, the Eternal became simply a natural—though to be sure, *supernatural*—phenomenon of the aggregation of all the world's unused life energy, and out of that tremendous synergy comes a sentience, a thing unto its own that we call God.

My lifelong concept of a Santa-God, an old white man with a

long white beard pouring over manila folders holding each soul's permanent record, was immediately dashed to pieces. It became clear to me that the Creator was more about physics than theology. In such light, I understood that there was no karma, for nature absolutely abhors paperwork, and no such thing as destiny, or fate, beyond returning to the One. There was also no heaven and no hell; those were just the dreams of a child. What I did know for sure was that I no longer had *belief* in the Eternal, I *knew* it as fact. And while I still seriously doubt that God can be petitioned for favors, I think it entirely appropriate to say thank you.

I don't know how long the vision lasted, but after it slowly faded, I walked back to our campsite mystified. I caught up with Pug at chow, and it was like I'd only been gone for a minute or two.

I didn't dare say a word to him about what I had seen. The whole world might think you're crazy, but you don't want your best friend to think so too.

———

When we got back from the field, Pug's transfer was already there waiting for him, and in a matter of weeks he was gone.

Though he wrote to me several times, I am ashamed to say I didn't respond to any of his letters. The truth was, I still had a few months to go on my tour and I had started to feel like an anchor holding Pug back, so I decided to cut him loose. It seemed best at the time to be cruel to be kind. But to this day, one of my single greatest regrets is not having the emotional maturity to stay in touch with Pug. He was the best friend I ever had, and remaining detached from him was a mistake. It would take a weight off my shoulders if only I could apologize to him, but as it stands, I will likely be forever mired in unextinguishable regret.

*October 31, 1985*

*Wie gehts, Harvey,*

*Happy Birthday! This is just your yearly reminder that you'll always be older than me. Did you go out and party? I hope you got some cake . . . or even a piece of hair pie. Is it as cold there? 'Cause it's colder than a witch's tit here. But in just a few months, I'll be seeing you before you have time to run and hide. So strap on your helmet—we're going to do some serious partying together. I'm going to bring some good shit back home with me, a little taste of Afghanistan, via Deutschland, if you know what I mean—and I think you do.*

*Do you remember when I told you about the CID, the Army's secret police? They're the lowest of the low disguised in civilian clothing. They busted a guy in the room next to me for being a homosexual. Thing is, we all knew this guy was gay, but he kept to himself and it was no big deal. Everybody's got their quirks and at least he wasn't a jerk or anything. He did his job, and I never heard about him getting into trouble—not like some people you might know.*

*Apparently he was "caught in the act" with a CID agent. Story is, they met at a bar, went back to his room, and somewhere in the middle of screwing each other the undercover piece of shit busts him. And I don't know who was doing who or for how long. But supposedly they were having sex so at the very least they should both be guilty.*

*So now that poor bastard is on his way to Leavenworth, and all because the UCMJ has an issue with his sexual orientation. How is it any of their business? It's not like he was blowing guys while they were supposed to be on guard. The guy totally kept to himself—and now he's going to do hard time before they drop him like a bad habit with a dishonorable. He'll have six years in the service (not to mention jail time) and nothing to show for it but a DD. And when they finally let him out, he'll be broke without any benefits. Ain't that a bitch?*

*And a fucking dishonorable can ruin a person's life. Every time he fills out a job application, they'll want to know if he served and what kind of a discharge he got. With a few asshole exceptions, everyone here feels that what happened was wrong.*

*Dennis, the specialist from my last unit is gay, and he's totally cool. He thought it was a secret, but everybody knew—and no one cared (especially the guys fucking his wife). I think my old sergeant major at headquarters might be gay, and I still respect the hell out of him. And I know one of the sergeants here is gay and he's a complete moron. He's in charge of my buddy's platoon and I really dislike him. So being a homo doesn't mean squat beyond the fact that they're not competing for the same women that the rest of us are. So where's the problem? I tell ya, sometimes (okay, most of the time) this Army is really fucked up.*

*I took my German girlfriend to Munich last weekend. You can get really good deals on hotel rooms after Octoberfest. It was a little weird, though. Usually I go alone and I think the locals were looking at us differently. Maybe it was my paranoia, but I don't think they like seeing an American GI with one of their girls. We went to a bar and actually got snubbed—they didn't want to serve us. I was set to make a scene, but I had Claudine on my arm and just let it ride. They can go fuck themselves cause I'm busy enjoying their women! How do you like me now?*

*You know what also really bugs me? And I'm only going to say this one more time. You won't give me the respect, the trust, to send me what I asked you. And that's really shitty. I might forgive, but I will never forget. My disappointment in you is profound. It reminds me of the curse laid upon Finn MacCool for refusing to help when it was in his capacity to do so.*

*I'm going back to the field in a month and it would be really appreciated if you sent a package before then . . . but I'm not holding my breath anymore. And think real hard before ever asking me for a favor.*

*Pissed off but still your brother,*

*Adam.*

# TWENTY-SIX

## Bagged Insubordination

When we were in the field, we often wouldn't have the luxury to bathe for weeks on end, and that's when I was glad it was cold—real cold. We couldn't smell each other as much that way.

We slept in small drab-green canvas tents that usually had a diesel-fed heater by the front corner, where it vented through a grommeted opening in the side. Even when that heater glowed a menacing cherry red, the ambient temperature in the tent was still about twenty below, which was better than forty below without the heater or minus sixty in the exposed wind. Nighttime was the scariest, when the sensory deprivation of the dark only highlighted the sense of being alone against the overwhelming elements.

Our cold-weather sleeping bags rested on small collapsible wooden cots, and there was one simple rule to live through the night in sub-zero temperatures: get naked or die. The problem was, it was far too cold in the tent to undress; the bags themselves were freezing, and by the end of the day we were all exhausted. But if you climbed into your bag fully clothed and fell asleep, which was so tempting to do, it was guaranteed you'd overheat in your sleep and begin to perspire, and that's the cold kiss of death. When the fabric of your clothes and sleeping bag become damp from sweat, they no longer insulate you from the intense cold outside and instead become frozen-solid conductors.

The trick was to entirely disrobe once inside the zipped-up bag, boots and all. Every article of clothing then got folded and placed at the bottom of the bag in the exact same spot every time. That way you'd have warm clothes in the morning and know where everything was even if you couldn't see it. But taking off your clothes in a frigid bag can be shockingly painful. So in the completely dark sack, just barely big enough to hold you and all your clothes, you'd do the cricket: rub your arms and legs together like you were trying to start a fire with your limbs, and the moment you felt the slightest blush of warmth you stopped—because we all knew the dangers of overheating. As the temperatures continued to drop, it would be necessary to repeat the cricket, as needed, throughout the night.

It was temping to keep the bag zipped all the way up, but there's too much moisture in one's breath, not to mention that after a few minutes one's own stink becomes overwhelming. To work around that, the Army issued snug-fitting, drab-green sleeping caps that fastened underneath the chin with padded velcro straps. I would sleep wearing my nightcap, with only my eyebrows to lower lip protruding from the sleeping bag, looking like a giant pickle with a face.

As my platoon's advance party, I alone carried my own smaller tent with a cot, though no heater, and often slept by myself. But when we were between fire missions, the whole Alpha company bivouacked together and I had the privilege of bunking with the rest of my gun's crew—though that had its own drawbacks.

It was cold enough that if the 8-inch's motor wasn't idled every two hours, the engine block would freeze solid, so we all had a shift to pull. Inevitably, four in the morning would be my turn to crank up the howitzer, AKA "the piece," and I hated that more than anything. We had to be up at five, so getting dressed at four to go out to start the engine for fifteen minutes killed what should have been my last hour of sleep.

Early one morning, during a winter reforger, I decided to stay put in my bag.

"Harris," yelled Sgt. Grimes shortly after four. "It's your turn! Get up and start the block."

*Maybe if I play dead he'll just leave me alone*, I fantasized.

Again, "Harris! Get up!" And again I didn't budge.

It was so damn cold that Sgt. Grimes didn't want to get out of his sleeping bag to make me get out of mine.

"No," I finally answered. There was a long pause. We were, after all, in a tent with five other guys and I was clearly being insubordinate.

"Harris," he finally said, "if you don't get out of your sleeping bag right this minute and start the piece, then as the Lord is my witness, so help me God, I will see you court martialed!"

That made me smile. I had already been busted lower than the rank I started off with. Let them tear off my E-2 stripes. Fuck those stupid mosquito wings, fuck Sgt. Grimes, and fuck the howitzer.

"Sgt. Grimes, you can court martial me if you want, but you're going to have drag me there in this here fart sack 'cause I'm not getting out," I swore in return.

By now we were all awake. The campsite around us was beginning to stir, and I thought I heard snickering from one of the neighboring tents. It was just about time to get up, but I tucked my head back into my sleeping bag and zipped it up. Rolling over, I presented my back to Sgt. Grimes and turned on my little handheld Braun electric razor, pretending to be shaving.

— — —

Sgt. Grimes had been a corrections officer in the CCF, Charlie's Chicken Farm—where bad soldiers were sent for reformation—and he saw himself as a real tough guy. He wasn't very big but he had a bulldog's disposition and a tremendous need to exert his authority. A few months prior he had been reassigned to our gun battery as my new platoon sergeant. When he reported for duty late the first night, he found the CQ, charge of quarters, deserted. Just down the hall from the CQ's desk was the rec room, whose back door was open and through which Sgt. Grimes glimpsed the fight that was occurring out back.

What he saw was a big guy on the ground with a little guy on top of him. When the little guy on top started beating the big guy's head into the pavement, the spectators, the CQ amongst them, stepped in to pull the little guy off. It took all of them to do so while the big guy was helped up, then limped off with a couple of his buddies. Enough fights broke out that it didn't make sense to stop one until there was a clear winner; tensions would just fester and brew otherwise.

It's probably no surprise that the little guy they pulled off was me. That was Sgt. Grimes's introduction to the Death Dealers.

Fifth grade is when I decided we human beings shouldn't pummel each other with our fists, and I swore off fighting for good. But in the Army, there are those who see fighting as a form of entertainment—and those who will entertain such foolish notions. I was one of the latter. I never threw the first punch, but the last one was always mine, which helped to limit my liability.

The morning after a fight, I had a little ritual I carried with me from childhood. I would ask my friends, usually while we stood in the chow line for breakfast, to point out the guy I had fought the night before. As always they would signal toward some big son of a bitch. I never did fight guys my own size; it just didn't seem fair. "Oh, shit" I'd say to myself as I'd walk over to him, trying to look up into his eyes in a menacing manner.

"Are we done?" I'd demand. "'Cause if not, I'm finishing it here and now."

I wasn't messing around. Inside I would be scared, but I knew as soon as the punches got thrown that feeling would disappear. And that as soon as I knocked him down, he wouldn't look so big.

Staring down at the crazy little confrontational Jew, they'd all say the same thing: "No, we're cool."

And that would be that. Nobody ever wanted second helpings. That way there were no hard feelings, I didn't have to worry about looking over my shoulder, and I never carried a grudge. That was my uniform code of military justice, but Sgt. Grimes was straight by the book and took a dim view of any deviation. To him things were black and white, and respect came with the stripes. But respect should never be given without just cause, and if respect is not mutual then it's hollow. Truth be told, respect is earned, and some make you earn it harder than others.

— — —

Sgt. Grimes and I got through our early-morning spat in the tent by ignoring each other. But our relations stayed as chilly as the weather, and having called his bluff earlier, I kept pushing his buttons throughout the day by paying him no mind. I was tired of living outside like an animal, tired of wearing green, and really tired of taking orders. Sgt. Grimes was just plain tired of me and my bullshit. By the end of the day, just as we had pulled into position, the sergeant reached his boiling point.

"Alright, Harris, you and me around the piece—NOW."

Once we were on the other side of that humongous machine, we would be hidden from the rest of the battery. It wasn't until we were halfway around, trudging through the snow side by side, that I realized he meant to fight me. Sometimes I'm a little slow on the uptake, but the rest of the guys figured it out faster and were following close behind so they wouldn't miss the show. Sgt. Grimes looked so damn grim. I smiled and put my arm around him, hugging him close to me.

"Sgt. Grimes, I like you—I don't want to have to hurt you." I could not have been more genuine or honest in that moment.

Sgt. Grimes stopped dead in his tracks and looked at me hard in the eyes with resentful anger. Then, common sense came flooding back to him. "Dammit, Harris!" He flung my arm off from around his shoulders and stomped away, cursing as he went.

Maybe he'd had a moment of clarity, or perhaps he remembered the first time he saw me. Either way, he wasn't stupid enough to let his anger take him to the same place as that guy whose head I'd been pounding that night months earlier.

Funny enough, my relationship with Sgt. Grimes improved a little after that. I took pity on him and he wasn't foolish enough to ever threaten me again with physical violence. He did, however, want to lock me up.

*Dear Harvey,*

*Hey, how was your Turkey Day? My buddy invited me to Thanksgiving at his fiancée's place. It was German/Puerto Rican style done with love. Have you ever had stuffing with smoked oysters? That's some good eating. We had a large roasted chicken because turkeys aren't available in the local stores, but chickens taste different over here—like the ones off the farm. It was, without a doubt, the best meal I've had since I've been in the service.*

*Really, the only reason I was invited is because I have a girlfriend—it was for couples only. I met Claudine a few months ago. She lives right across from the billets and saw me jump the wall one night. I felt her eyes on me and went over to make sure she wasn't going to call the cops . . . and we've been seeing each other since. It's not a perfect match but she keeps me out of trouble. She's a sweet girl and I'd hate to think I'm using her, but I think we're both using each other. She wants me to stay but what would I do here? I don't speak the language or have a degree.*

*I just need as much distance from the Army as I can get right now. This place is full of bored guys with too much testosterone and at times it's a powder keg. I keep a full set of BDUs at Claudine's place, and if we get called out it's easy enough for me to run across the street and jump back over the wall into the Post. In the chaos of an alert no one would ever notice and, like jail, no one stops you from breaking in.*

*And like jail, no one's safe. My buddy Big Ben got the crap beaten out of him. Ben was the first guy in my room when I transferred to the Death Dealers to say "How you doin?" and shake* my hand. It didn't hurt that I had a bottle of tequila in my other hand. Ben's from Jersey, and he talks like Rocky Balboa. Really he's just a big teddy bear, not naturally violent or aggressive. To fit in with the Death Dealers, he acts like a tough guy. It doesn't really work, but no one has the heart to call him on it. Not until last night.

Ben always talks about becoming a boxer once he gets out of the service, but he's not going anywhere. He doesn't have a whole lot of options and he's not the sharpest tool. With just a GED I'm pretty sure he's a lifer. Ben's folks sent him his eight-track player from out of his car, along with the speakers, like he asked. He's got it wired up to one of those big 6-volt batteries, and he shadowboxes with Springsteen blaring super loud.

It's funny 'cause Ben's got these hand wraps and sparring gloves, but he's never stepped inside a boxing ring—and now I know why. But he really goes at it in front of a mirror he's got on the inside of his locker, and it bugs the shit out one of my other roommates—a new guy I call the Dickhead. He gets all bent out of shape 'cause he doesn't like rock 'n' roll and he doesn't like Ben getting all sweaty in our room. Basically, he doesn't like Ben.

So the two of them got into it last night. They were both liquored up and it did not go well for Ben. I wasn't around to see it happen, but I saw him right after and his face was brutalized. He had so many lumps and bumps that his head was misshapen—like something out of a comic book—with a busted lip, bloody nose, and a black eye to boot.

Still he's a stand-up guy, so when our sergeant saw him next morning and asked what happened, Ben says, "Hey, yo, Sarge . . . guess I musta tripped and fell down the stairs or sumthin . . ." And because Ben isn't a troublemaker, our sergeant just let it go at that, *which is cool. But I still feel bad for Ben—he didn't deserve what he got.*

*So I've got to get out of here, but I'm still trying to figure out what I'm going to do and where I'm going to go. I've got to go back to school. It's the whole reason I'm putting up with this crap in the first place. Maybe I should study psychology. With all the crazy shit I've seen, I've got a certain perspective. Perhaps I can make something of my time here. Like reaping a harvest sown from these seeds of pain and misery.*

*I don't want to go through another Midwestern winter, and if I get out after spring semester has started—then what? I'll have to find work . . . and as long as the pizza place needs a delivery guy who can operate a fifty-cal, then I'm in like Flynn. Do I tell them I was the fuse setter for the Death Dealers? 'Cause I think it'll scare them away.*

*I got a letter from Mum the other day and it sounds like I'll make it home for Shana's bat mitzvah. I could really use a dose of home. It'll be nice to be around family, and I could probably use a little religion too. That's the stuff of dreams. Are we going to put the folks up on chairs? You know the rules: three short guys and one tall guy per chair . . . where do we find two tall guys?!*

*I made a countdown calendar . . . it's in the shape of an airplane. I call it "The Freedom Bird." It marks my last 50 daze . . . I am officially a double-digit midget. All I've got to do to ETS—Exit This Shithole—is to keep my head down. Wish me luck.*

*Like the Count of Monte Cristo—I wait, I hope, and I plot my escape. See you a little into the new year.*

*Love, Adam.*

# TWENTY-SEVEN

## Rabbit Playing Possum

As far as I knew, police brutality laws did not exist the years I was in Germany. My first roommate warned me that the German Polizei had carte blanche and no love for American servicemen. He told me a cautionary tale of a bunch of Kraut cops shattering a GI's nose and left eye socket, then throwing his body into the street in front of a US post and speeding away. The guy lost that eye—and there was no recourse. They saw Uncle Sam as a houseguest with a gun obsession who just wouldn't leave, so diplomacy was the way to maintain an amicable relationship with our German host. If that meant turning a blind eye to a couple of beaten-up GIs, then so be it.

It was only forty years prior that the world-conquering "Deutschland, Uber Alles" was forced to surrender to the Allies, and the American servicemen were largely still tolerated with the same disdain as they were then. Germany was trying to separate its present from its past, waiting for what was swept underneath the rug to die, and we were in the way. It probably didn't help that we drank their beer, screwed their women, and acted like Germany was part of our disposable American kingdom.

Ancient Germany was a Neanderthal stronghold, and in my

observation, that primitive race didn't die off; they merely assimi-
lated. In the local farmers' markets, little old ladies with pronounced brows and soup-can ankles would elbow me out of the
way to get to the produce. The most useless German phrase I
ever learned was *"Entschuldigung,"* meaning "Excuse me." I was
the only one I ever heard say it. And when I used public trans-
portation, I couldn't help but look at the old people riding the bus
and wonder, *How did you survive? How many of my family did you help
kill so you could sit there looking at me so smug?* The oft-quoted statistic
of "6.2 million" was not only terribly conservative; I knew that at
least half the German Jews murdered during WWII were killed in
their homes, in the streets of their neighborhoods, and in their
places of employment, and I couldn't get that out of my head.

During maneuvers, we would roll through little German
towns in our howitzers and APCs, ripping up the cobblestone
roads, squashing the occasional car, and once in a while taking
out the corner of a building with an errant gun barrel. I admit,
that's when I liked to ride on top of our howitzer. With my M-16
rifle in my hand and a cigar in my mouth, I felt very much like
the victor in an occupied country saying, "How do you like me
now?"

François Kaserne was one of more than half a dozen US
posts in the small town of Hanau, which was bordered by the
Kinzig River and surrounded by the Black Forest. In the town's
center—not far from the statue of a couple of local boys, the
Brothers Grimm—was a circular, multi-tiered shopping center. At
night the shops closed down and the restaurants occupying the
basement floor turned into taverns. There certainly seemed to be
a lot of single women in Hanau, and we GIs would head for those
pubs in droves.

After one such night of flirting, drinking, and dancing until the bars closed, a bunch of us headed up to the third floor to a little "taco" wagon run by a young Sri Lankan couple. They didn't speak a word of Spanish, and they sure as hell had never been to Mexico, but that pseudo Tex-Mex was ambrosia to our inebriated states. More like a split, soft samosa of generous size topped with an assortment of sauces ranging from white-hot to fire-breathing orange, the food was tongue numbingly spicy, but so good that it was hard to stop—and perfect to soak up all that alcohol in our systems.

Happily drunk with a full belly, my friends and I went down to the ground floor to catch a taxi back to the post. The dull yellow street lights seemed to cheapen the pristine December air, tinting my orange East End Ruffians jacket a deep blood-red. The cold wind felt good against the sweat on my forehead from the spicy food, and I was almost disappointed that our cab showed up so fast.

My buddies slid in first, and as I was about to climb in after them, I was suddenly grabbed from behind with tremendous force and yanked out of the cab. I was flung hard to the ground before I could even react—and then the stomping began.

I briefly saw the uniforms and knew it was the Polizei. They kicked me like a bunch of truculent school children with cast-iron boots playing "kick the can," only it felt more like they were trying to *kill* the can. I'd been on the bottom of enough loose rugby rucks to know what to do: cover my face with one hand, my groin with the other, and keep my eyes toward the ground. There was no appealing to my attackers—there was only doing what I could to stay alive.

As my friends sat helpless in the cab with front-row seats, I used every ounce of willpower to remain still until they stopped. If I moved even reflexively from the pain of being struck, they would kick me all the more. Same if I let a sound or moan escape. Only when I was a motionless ragdoll did the beating stop—and even then they made sure I wasn't acting.

In those untenable minutes, as my body was being viciously pummeled on all sides, my mind sought solace and I entered a separate reality where everything was cozy and peaceful. Whether that's what happens to gazelles when a lion has them by the throat, or I was simply knocked out, I'll never know. But I remember feeling at tremendous ease, warm and comfortably tired, lying on a soft mound of grass. It felt like I was ever so gently sinking into it, and everything seemed serene and safe. It was profoundly peaceful, and I wanted to stay there—and then I noticed a severed arm lying next to me. Instead of being weird or scary, though, it was natural in a familiar sort of way.

I remember thinking I had seen the arm before, as if in a dream I'd had long ago and could no longer recall. It was as still as I was in the blood-soaked grass. And I could smell the blood. It reminded me of my mother's cast-iron pan, with its underlying sweetness, and made my nose burn. The hand of the detached arm was holding on to a battle axe, defiantly refusing to relinquish it even in death. A soiled and torn wreath lay discarded on the ground near us, with the faded and stained words "All for One" scrolled across it.

I could taste the blood now, and there was a cacophonous roar in my ears—like an ocean of shouting voices and bleating car horns. I felt the grit of the road as they pulled me by my arms across the street and bumped me up over the curb on the other side. I was then banged and bruised down three flights of stairs as they dragged me to one of the basement taverns below.

In a spectacular demonstration of strength and superiority, four of the Polizei lifted me up with the sole purpose of throwing me back down to the cold stone floor at the tavern owner's feet. He nudged my shoulder with a scornful shoe tip.

"I said he was wearing a red jacket—not orange!" the proprietor spat out in his guttural German.

Imagine that, the German cops had kicked the shit out of me before they figured out they had the wrong guy. To make up for their blunder, the four guys who had picked me up just to throw me back down repeated the act. After tossing me to the ground again, the uniformed prick-in-charge said to me, "Be careful, next time you won't be so lucky. Now GO!"

That was my cue.

I sprang up like a rabbit playing possum and ran up the three flights of stairs to the top floor. Stopping at the landing, I could see the Polizei standing below.

"*Essen mein scheisser, ashlockes,*" I shouted down to them as loud as I could. *Eat my shit, assholes.*

With their undivided attention, I spun around and pulled my pants down so they could get a good look at their reflections from where the sun doesn't shine. Then I pulled up my pants and sprinted like I had just offended the Angel of Death—and like my life depended upon it, which it did. The Polizei weren't known for their sense of humor. I knew if they caught me again, there'd be no running away a second time.

*Hey Harvey,*

*You'll never guess who I ran into . . . Steve Pinkney! He looks just the same as he did in high school—he joined up about five months ago. Do you ever hear from his brother Ryan? I was coming out of a Frankfurt whorehouse and he was on his way in. That's how we do it after we get back from the field. After a month of sobriety with Rosy Palmer and her Five Sisters as our only intimate release, we all head for the Frankfurt Red-Light District.*

*Some guys start in the bars and never leave; some guys spend fifteen minutes in the brothel and then go home. Actually, they're not brothels, and not like the Turkish whorehouse, but more like a big apartment complex, where every unit is just a small room with a bed and a sink. The girls wait by their open doors if they're available and there is every kind of woman—from tall to short and fat to skinny. If there's money in your pocket, you're going to spend it for sure. And in my biz there's no point in saving for a rainy day.*

*So the night I saw Steve, I was going through one of the complexes and this guy comes out of a room ahead of me with a big grin. It looks like he just had the best time of his life. He sees me and says I've got to try out this chick. And then he puts fifty marks in my hand and says he'll even pay for it. So I figure, what the fuck? Inside the room is this absolutely beautiful woman. With a really warm smile, she brings me into her room and closes the door. I'm still wondering what's the catch—'cause if she's got a dick then I want the money back.*

*We sat down on the bed and the first thing she did was tell me that she lost one of her breasts. That it was cut off with a machete. She's a refugee from Rwanda and is settling in Frankfurt to escape the violence in Africa. Then she opened her blouse and showed me her chest—one absolutely perfect breast and the other one missing. She took my hand and placed it over the scar. It was bigger than my palm. And then she kissed me.*

*I felt so many emotions all at once. She opened up my shirt and started kissing her way down. Hot damn, I almost lost it right then and there. And she was incredible in bed. It's hard to describe, but it was like she wanted it more than I did—and that really turned me on. When we were done, she dressed me and sent me on my way with one last kiss. I know you should never kiss a hooker on the mouth, but I also knew they had whiskey at the bar I was heading to.*

*It's funny but I can't seem to get her out of my head. She was truth, beauty, and tragedy all rolled into one. She haunts me . . . the courage she has to endure so much, all to have a normal life. It's not so much to ask, is it? Sometimes I wish I was more than just a private. I should have stayed in school. How do you help someone with limited resources when you're so limited yourself? I need to be more, to be able to do more. There are times that I feel so small.*

*I was reading about self-actualization in a magazine, how through the power of positive thinking you can attain your goals. Last night my buddy Slim and I were in a small bar and that Billy Idol song about dancing with yourself came on. Slim wanted to dance but none of the women were interested in him—he was pretty drunk. To tell the truth, we both were. The old Elvis line came to mind and I yelled, "If you can't find a partner, use a wooden chair!" So Slim grabs a barstool and starts swinging it around like he's doing the Lindy Hop with it, and the Korean bar owners rushed out to grab him before he smacked somebody.*

*I tried to run interference for him, but I was laughing too hard* and they 86'd the both of us. It's not like we don't have a ton of other bars to go to in "Drown Your Sorrows Deutschland" . . . hell, alcohol is the one thing that keeps the German economy running. And booze might not be the best for the body—but stress kills too.

Freaking A-right it ain't easy being green and as usual my head's pounding. But seventeen hundred is right around the corner. Cheers!

Signing off for now, and remember, don't do what I do—do what I say.

Love, Adam.

# TWENTY-EIGHT

---

## Pickle in the Stew

I met only one other Jew during my entire time in the Army—a new section chief replacing the departing Sergeant DumbOx. I was thrilled at first, thinking anyone would be better than that moron, and maybe I'd have the chance to go to services for the High Holy Days at the very least. But our new sergeant was as close to a Jewish Uncle Ben as there is in my culture, and he took it upon himself to drum the Jew out of me—for my own good.

After his first formation with Alpha Battery, he called me aside. "Harris, I understand you're Jewish."

"Yes, Sergeant," I replied, not knowing where this was going, but ever optimistic.

"Well, I've been a Jew longer than you've been alive . . . and I've been in this man's Army for a hell of a lot longer than you ever will be." He paused briefly to narrow his stare on me. "In case you didn't notice, this is a Christian military, protecting a Christian country. And you, Harris, you stick out like a pickle in the stew."

*I hate these American Army colloquialisms*, I thought. *Is being the pickle a good thing or bad?* It sounded Polish to me, and I knew too much about the Poles. My grandfather's second wife was Polish. She was a quiet, almost shy woman, who seemed to be permanently shell-shocked after being the only survivor of her tiny village. I asked her once about the Polish, curious about where she came from. "Fuck the Poles," was all she said. I had never heard her swear before, or after, and that was as far as my line of questioning went.

The new section chief went on to explain that it wouldn't do me, or my fellow soldiers, any good by holding on to a separate culture. I needed to get that "Jew chip" off my shoulder and get in step with everyone else.

The whole time he blabbed on I watched his mouth, a thread of scarlet below a bulbous nose, and let the words fall dumb off my ears. He was a looming man, with a high forehead overshadowing dull, beady eyes. All I could think was how ironic it was to be under the thumb of a self-hating Jew in Germany. That's when I decided his nickname would be Sergeant Schicklgruber. And seeing as he was the only other Jew I ever came across, he couldn't have known that many others himself. Maybe there had been one who had indoctrinated him like he was trying to do to me. Little did he know he'd have better luck squeezing blood out of a matzo.

As a member of the "tribe," Schicklgruber took a special interest in me—and my apparent lack of discipline. And it wasn't really about me. I was just an easy target. He thought if he came in taking names and kicking ass that he'd be promoted up and out that much faster. The gun battery wasn't the kind of place anyone wanted to stick around for long.

It had been less than a week since my run-in with the Polizei at the mall, and apparently some fat Norwegian rat looking to ingratiate himself told Sergeant Schicklgruber that I had been beaten up by the German cops. That's all it took for son-of-a-bitch Schicklgruber to seize his opportunity to have leverage against me.

A funny result of the political warming brought on by Gorbachev was the increasing feeling that the US military had outstayed their welcome in Germany. There had also been a spate of altercations between GIs and the Polizei, and we had been warned that an Article 15 would be waiting for any serviceman who got in trouble with them—regardless of the cause. Improving our image outside the gates was the order of the day, and an American soldier being arrested by the Polizei made for bad politics. It also made the guys at the top sweat the kind of shit that grows as it rolls downhill—especially when it showed up in the local papers, fueling anti-American sentiment.

After morning PT, Schticklgruber came into my room and ordered the rest of my roommates out. Big Ben had to finish dressing in the hall. Then he told me to take off my uniform. I looked at him like he was crazy and flatly refused. I didn't know what his game was, but I knew what he would find. So he threatened me with "insubordination" on top of "damaging government property"—that property being me.

"Harris," he stated matter of factly. "The Article is already written up. All I have to do is sign this and you're gone."

I clenched my jaw and reluctantly held my tongue. Then I took off my shirt and let my pants fall to the ground—that was good enough for show and tell. Almost all the marks had a similar shape, as the Polizei who kicked the crap out of me all wore the same type of steel-toed boots. The discoloration—ellipses of black and blue, highlighted with deep purple, angry red, and sickly yellow—depended on the intensity of the kick and the depth of the injured flesh. It was quite dramatic and reminded me of a spotted character in a favorite Dr. Seuss book. Schicklgruber appeared taken aback by what he saw. I hoped he felt ashamed of himself for adding insult to injury.

But he wasn't going to let me off the hook that easy—oh no. Our battery's Christmas party was coming up and Sgt. Schicklgruber had the audacity to tell me that the function was mandatory. Without missing a beat I told him straight off that it was a religious affair and I had no intention of going. Then he mentioned Sgt. Grimes's strong desire to send me to CCF, the correctional facility. An Article 15 would land me there for sure. Even if he was full of shit, I knew he could do his best to make my life even more miserable than it already was. I was getting short, but I wasn't untouchable.

He walked away from me with a stiff warning: "Show up at the Christmas function or get ready for the correctional facility."

We had all heard stories of the CCF—how it made Basic Training look like a tea party. But it wasn't the physical aspect that scared me. It was the thought of being locked in a cage, and of losing myself, that sent a shiver through my core. And so I swallowed my pride. The party was from 18:00 to 21:00, and I resigned myself to a quick appearance before getting the hell out of there.

That evening the billets were empty, as if everybody was already at the party. I reluctantly headed over to the event about an hour after sunset, hoping I could slip in and out of the shadows without too many people seeing me. As I got to the top of the steps, the first person I ran into was the snitch Fat Finn and we locked eyes. Doing anything to get ahead, he was dressed as Santa Claus—and knowing I had been ordered to show up, he was waiting for me, looking to rub it in.

That was my breaking point. I refused to give him the satisfaction, and no amount of threats could have made me go any further.

As I turned away, begging my legs to move faster than they could without running, my eye caught Sergeant Schicklgruber's. He had that "I smell shit" look all over his face as he watched me do my U-turn without passing over the threshold, then head straight across the street to the bar for a night of Johnnie Walker Red, Johnnie Walker Black, and trouble.

Bright and early the next day, right after morning formation, I was called into the top sgt.'s office. If Schicklgruber couldn't turn me then he was going to get rid of me—but his bid to send me to CCF must have fallen through. Instead, I was transferred to the front gate to work with the Unit Police, with whom I had a very uneasy truce.

Schicklgruber came up to me after I had been dismissed and explained that we had to be one unit, one All-American team, and I clearly couldn't do that. So he was going "to isolate me, for my own good." I believe he was sincerely twisted.

The U.P.s knew better than to mess with me. Luckily, there were only three of them, and they all gave me the cold shoulder. I spent the entire day standing outside the shack, manning the outer gate, waving traffic on and watching it go by. Beyond the plan, I preferred the cold and my own company to a hot box full of assholes and their recycled air.

At the end of the day, I purposefully remained at the outgoing gate to hold the most critical position—allowing cars off the base. Regardless of rank, the point guard manning the outer gate held sway over all. It was that guard's job to determine if the strasse was safe enough for vehicles to leave the post, and it didn't look safe enough to me, so no one was going anywhere. I may have been a lowly private, but I was going to make sure they would regret putting me with the U.P.s. It's what I had been waiting for.

"That's MY fucking post," said the ranking U.P. as he came

out of the shed. "Seventeen-hundred . . . step aside you gun bunny reject."

I stepped toward him. "I'm gonna throw you back through that fucking window if you don't back the fuck off me right fucking now." I was grinning widely. One way or another I was going to have some fun.

"It's—" he started.

"It's time you got the fuck out of my face," I finished. "Get back into your shack or I'll kick your ass up and down this street. GO!"

He chose health and retreated back into the shed, leaving me alone to deal with the flood of outgoing traffic, already growing impatient. I may have been a puny oarsman, but I was going to rock that big boat—and they would know who HardWay Harris was.

I had taken possession of that narrow pass with a plan, and it was coming to fruition. Officers don't like to ask permission from, let alone wait for, some lowly enlisted—it's like a reversal of their whole Army hierarchy. I was really only curious to see how long I could shut the place down, and what would happen in case of an alert. Where would all those officers and their little German cars go if a couple of big ol' five-ton trucks came barreling through? This was my one chance to roll that ball of shit back uphill. Somebody was going to get in big trouble for putting an idiot like me in charge of the front gate during rush hour.

Respect amongst officers is like honor amongst thieves, and in the melee of blaring horns and authoritarian shouting, several junior officers got out of their cars just to yell at each other—trying desperately to establish a pecking order. One very upset major, who was missing his dinner, thought if he got up in my face he could scare me off. He didn't know me and he didn't know that I read the manual, but I assured him that until properly relieved I was in charge of the front gate, and nobody was going anywhere until I said so.

"Now you get back in your fucking car, major," I whispered slowly, stepping right up to him and pinning his left foot beneath my right as I leaned in close. With a sudden backwards jerk he flinched free, pausing to glance up from his scuffed shoe at me before returning to his car. I was finally dismissed by the sergeant major himself, red-face boiling angry with a voice shrill as a tea kettle. He stripped me of my U.P. shoulder bands and vest, then ordered me back to my own barracks—like sending Brer Rabbit back to the patch.

By then, the bottleneck had turned the front of the post into a rock-solid traffic jam. It took hours to sort out because as time went on and tempers flared, officers kept trying to pull brass over one another just to get ahead—which only led to further bedlam. And the U.P.s? They just stood by like subservient fools.

The next day I was reassigned to a guard shack at the mouth of a deserted warehouse a mile away from the main post. The place was completely locked up with only a small shed out front that became my office. Nobody else worked there; no people or packages or traffic of any kind came through. I'd walk there hung-over in the morning, smoke cigars and read my books during the day, then head over to my girlfriend's in the afternoon. In their attempt to bury me, I was finally given the peace and quiet I had longed for. I didn't have to fuck with anybody and nobody had to fuck with me. I'll never know if someone took mercy on me, or if simply being abrasive had its own rewards.

*Happy New Year Harvey,*

*I hear you and Jerome are venturing up to the Great White North. Do me a favor: when you see Grandma don't tell her I was counting her out. I guess there's more stuffing in that old bird than I thought! Maybe I was being a little pessimistic—seeing the darker side is a byproduct of this military lifestyle. It's good to be wrong sometimes . . . and some people might even say I make a habit of it.*

*I must admit, I think the Army is changing me. Most of the time when I go out on the town I'm looking for women. But once in a while I get this feeling like I just want to get into trouble. It's weird to say, but there are times that I feel distinctly violent. I don't remember feeling like this before. Sometimes it's like there's a smell in the air that triggers it . . . like how it smells before a storm. And I know I'm going to get into a fight. Hell, I go looking for one.*

*When the feeling hits, I have a little ritual: I order a shot of Johnnie Walker Red and a shot of Johnnie Walker Black. If you were to hear me order that in a bar, and you weren't my friend— your best bet would be to go drink somewhere else.*

*The other night at the disco, I noticed this couple dancing. This chick had an amazing ass, and I couldn't take my eyes off her. It's like her ass was hypnotizing me and, as I'm not shy, I let her know I noticed. I guess the guy she was dancing with didn't appreciate the competition. So he comes over to me and tells me to stop checking out his girl, and I tell him there's no law against looking. To make my point, I get right behind her and drop to my knees so her perfect ass is right in my face (if she had farted it would have been funny as hell). But I was just proving to this guy that he couldn't tell me what to do. I figured I was giving him no choice but to put up, or to* shut up, so he just left with his girl. And I've got to admit, I was disappointed—I really wanted to fight.

The Balinese say that to be balanced you have to have a foot in the good and a foot in the bad. I can't explain why I feel so aggressive sometimes, but maybe I'm still searching for my balance. Or maybe it's 'cause I take orders from assholes all day. Dickheads who treat me worse than furniture in a frat house. I hope, when I'm done over here, this thing runs its course. This is not who I want to be—but sometimes it's who I am.

And talking about a fucked-up world . . . our barracks has these old tacky and torn Christmas decorations strewn about—and a pathetic little plastic tree by the CQ desk they've still left up. I swear I'm going to set that shit on fire and call it a menorah. There was a "special" Christmas Mass (seemed like I was the only one who didn't go . . . what the fuck do they do in there?) and a Christmas Day dinner. There were even Christmas presents for us. All we had to do was sign for them at the CQ.

They couldn't give a shit that I don't celebrate Christmas, that it's not my religion and not my fucking holiday—so none of that hoopla was for me. And the gifts were just a bunch of stupid label machines, so motherfuckers can label their stupid shit. Man, I hate this time of year.

It's just that my entire world becomes so super saturated with their religious propaganda. We get one TV station, and from Thanksgiving till New Year's it's all Christmas-themed crap all the time. I swear if I hear "The Little Drummer Boy" one more rump-a-tump time, I'll kill someone. Jesus was never the King of Jerusalem, and when they reminisce about a "White Christmas," I know it doesn't have anything to do with missing the snow.

*I spoke to the chaplain when I first got here to François kaserne about making latkes for Hanukkah. I figured we eat potatoes every meal anyway . . . they're easy to make and I even volunteered my time. That afternoon I was called into my top sgt.'s office and was told to leave the chaplain alone. That's how the son of a bitch said no, and that's when I really started to fuck with him. Now he runs and ducks for cover every time he sees me.*

*Well, thanks for letting me rant . . . I feel better. I'll be seeing you soon and I'll bring back some souvenirs—unless I get nabbed at the border, and then you can always visit me in Leavenworth. At least you can get there by car! Otherwise, tell 'em Daddy Cool is on his way. See you soon,*

*Love, Adam.*

*PS: Do you know about Daddy Cool? Across the street is a chicken place where they fry pieces in a smoky paprika oil and the chicken comes out orange and super spicy—perfect with a cold beer. They have a video jukebox to watch while you're waiting and I usually play Sheila E. (man, what I wouldn't do for that woman), and the new Phil Collins' World of Confusion is a big favorite. But the other day I noticed a song called "Daddy Cool" by Boney M, and I was so intrigued by the band's name that I had to play it.*

*Boney M is made up of three sexy black chicks and this one crazy dancing guy. I mean nobody's got moves like this dude. And there's no special effects . . . just a hilarious contrast between this really freaky band. About halfway through the song, one of the singers starts moaning like she's having an orgasm, and the German audience obviously doesn't know what the hell they're watching or how to react. I mentioned them to Claudine and she gave a little shriek, went running into her room, and came back with her old record player (it's a compact one like Esther had when we were kids) and her first record ever. You guessed it—Boney M!*

# TWENTY-NINE

## Be a Sport

Kash was the ranking specialist in our room. His bed was by the entrance and he liked to keep the door closed, so whenever anyone knocked, they were met with "Speak" shouted so loudly that it echoed down the hallway behind them. If you didn't have balls big enough to sound off, you were left standing outside the door. What one wouldn't have guessed is that Kash was really a nice, quiet guy, an introvert who was comfortable playing the Army game—not to mention that as a neat freak, the absolute orderliness of the service suited him well.

Kash was one of the few "other" people I knew, meaning a person of unrecognized culture who didn't fit neatly into the classification system. You could be black, brown, or white in the Army and know you'd be well represented. That's how America likes to categorize its people. But outside of that color palette, like me, you checked the last box to the right—other.

Kash defied definition. He was tall, handsome, and deeply crème brûlée colored, with a square jaw and fine features, dark almond eyes, and jet-black hair. He never discussed his ethnicity or religion; it was something he was very guarded about. And it wasn't my place to ask. I always imagined his father from a Kashmir mountain, tall with raven hair, who had somehow met and wed his mother, a dark-caramel island girl from Bali with even darker eyes.

Kash and I naturally gravitated toward each other, maybe because he, too, carried a chip on his shoulder, and because he also knew what it felt like to be singled out, to not fit in through no fault of his own. Together we made the perfect odd couple.

Kash's appearance was beyond reproach. His uniform was always pressed and his boots always spit-shined—even in the field. He was lean and trim, which made me look shorter, more wrinkled, and boxier next to him. Even though our BDUs were identical, on him it looked like a custom-tailored suit, while mine looked like I found it in the Salvation Army bin. Where Kash moved with an innate sense of stealth and grace, I lumbered like an angry dwarf; where he was retentive, I was explosive. But what we had in common was that neither of us was a dummy or pushover, and our mutual respect came easily.

I never expected my lasting and indelible memory of Kash to be sliding down the Strasse on his turtle-shell of a radio.

We had been called out on alert and our five-ton was already rolling when the last of us jumped into the back. I carried the heaviest load—the ammo for the M-60, the classic all-purpose Army machine gun, plus, of course, my own weapon and rounds.

For such a light, fast-moving deployment, the M-60 was our biggest firepower. But without bullets it's just a hunk of metal, and it uses up a lot of ammo: over six hundred rounds per minute. Normally the M-60 rounds were divided amongst everyone in a platoon, but for the alerts, one guy alone was assigned to carry all the ammo for the machine gun—the guy who could run with it as fast as everyone else could without. And so, even with a limp, it had to be me. And if the machine gunner went down, that became my responsibility as well.

As our communications guy, Kash had the second heaviest hump with the radio: a suitcase-sized metal box strapped to him like a backpack. When the alert was issued, we had sheer minutes to be out of the sequestered cabin and into the back of the five-ton, its engine revving in anticipation. Due to our loads, Kash and I were always the last two onto the truck, already in motion as we roared toward the front gates. But during one deployment, the back lift-gate wasn't properly latched, and as we turned the corner onto the street it fell open. Kash had been leaning against it, so the bump that jarred the latch open bounced him right out of the back of the truck—at considerable speed. The guys who had seen this, myself included, started yelling at the driver to stop, but those damn five-tons are as loud as they are heavy, so our pleas fell on deaf ears.

The centrifugal force had flung Kash into the middle of late-night traffic, the weight of the radio holding him like a flipped turtle as he slid across the street behind us, arms and legs flailing in the air amid a shower of sparks shooting out from the radio beneath him. Again, we pounded on the front partition of the five-ton, but the driver gave no sign of slowing.

Then, like something out of a Monty Python skit, Kash popped up and started running after us. From my vantage point, I could see the look of grim determination on his face. He sprinted for two solid blocks until we were forced to slow down, at the one and only intersection that stood between us and the autobahn. In what must have been a tremendous burst of adrenaline, Kash defied logic and caught up to us. As traffic cleared and we roared forward, he leapt for the open back of the vehicle. I had positioned myself on the bumper, holding on to the side grip with my right hand. As Kash flew toward me, I caught his wrist with my left hand and hauled him aboard. Breathless, we took our seats and sat in awed silence, as if minor miracles like this were all part of the job.

— — —

One day after work, over a few beers, Kash leaned over and confided to me that he found women's feet sensual, sexy, and even stimulating. He had seen a catalogue with rubber models of women's feet and had thought of making an order, but was afraid it would make him a sexual deviant. I had never heard of a foot fetish before but it sounded harmless to me. I agreed that women's feet were beautiful, as was the rest of them, and if there were tits-and-ass guys, why not feet guys too?

I assured him there was no shame in that and suggested we go out. That night at the disco we spotted two girls sitting by themselves, one wearing canary-yellow pumps and the other bright orange stilettos. I didn't know the difference, but Kash sure did. I'd never seen him so happy; he looked like a Tex Avery wolf. And so we sat down and introduced ourselves. I'd like to say that the rest is history and gentlemen don't tell. But the truth is that Kash left with both those lovely ladies that night and I went home alone.

— — —

When Kash's time came to be transferred I was, as usual, on suspension and not allowed to leave the kaserne. The little honky-tonk on post was closed for repairs, so we decided to go out for his last night on the town. In the spirit of camaraderie, Kash, Roland, and Big Ben jumped the wall with me, which is the only way I got on and off post and why I knew the perfect spot. Kash went first, followed by Roland and then Ben. As I was about to jump, "Halt!" rang in my ears and I was instantly blinded by the glare of a sentry's flashlight. The way he spoke, however, without a glimmer of the voice of command, told me I was dealing with a newbie, and I had a sudden memory of a lesson I'd learned when I was about five years old.

There was an enormous German Shepard named Sport who occupied the enclosed front porch of our farmhouse. He was no-body's pet; he oversaw and overheard everything that came and went, and the only person he paid any mind to was my grandfather, a relationship that was akin to Cerberus and Charon. One night, I approached Sport with curiosity. His shoulder was as high as my own, and he turned with a growl that vibrated through my skeleton and froze me to the quick. With his head lowered over his bone, he bore his teeth while the hair on his back slowly rose. This instant ferocity with which he projected his menace, with absolute assur-ance of his abilities, was what was so terrifying. I never went near Sport after that and that suited him just fine.

With the same savage snarl Sport had taught me so many years ago, I turned toward the guard. My hand reached from the shadows for his flashlight and lowered it as I stepped toward him. Like with animals in the wild, certain sights and sounds tap into the primitive portion of our brains, short-circuiting everything else. My time was short, my buddies were waiting on the other side of the wall for me, and it was my last chance to have a drink with Kash. There was no way in hell that U.P. was going to stop me—plus I had to scare him silent; otherwise, we'd be busted when we got back.

"I know where you live," I growled with authority. "I know where you eat and where you sleep. If I ever hear one word about this, I'll be coming for you. No matter where you go, I'll find you. You didn't see anything. Nothing happened here. Do you under-stand?"

He gulped hard, nodded, and mouthed something.

"One word and I'll spill your brains with a claw-hammer. You got it?"

"Yes" came the spiritless answer.

I promptly jumped the wall without a look back.

———

We didn't return to our jumping spot at the wall until about 03:30, when the moon has already quit but the sun is too slow to follow. Silent as phantoms, we slipped back over into the kaserne. In the dampening mist, we came upon Sergeant Schicklgruber's Jeep parked behind the barracks and decided to do a little impromptu "detailing." As if practiced, we each took a position and set about tearing apart his vehicle.

Unable to puncture the tires, I let the air out of all four as Kash went about meticulously removing the windshield wipers, front and back. Roland, to my amazement, grabbed hold of the edge of the weather stripping around the driver's side window, peeled it off, then moved on to the next window. Big Ben was not to be outdone; he snapped the side mirrors off with a little leverage and a whole lot of body weight. How we didn't get caught in the process with the racket we were making is beyond me. It wasn't until just before reveille that the billet's back door opened up—and we froze.

Sgt. Lopez had been up all night manning the front desk and wanted some fresh air. Lopez had no love for Schicklgruber either, so when he spotted us, he quietly ushered us in. We crept silently up the stairs and made it into our room only minutes before wakeup. Not an hour later, as I was getting dressed, I peered out my window to the parked cars below and caught my first glimpse of the real Men in Black—black suits, shoes, ties, and sunglasses, one holding a black briefcase, going over the sergeant's fucked-up car and cataloguing all the damage.

What struck me most was that they didn't have the look or smell of military men, but rather the foul offspring of lawyers and accountants as they brushed for fingerprints, used measuring tapes, and took Polaroid pictures.

"Oh, shit," I said out loud. "We are so busted."

We had been too drunk to be careful as we tore up the sergeant's car, and those guys were certainly going to find all the evidence they needed. The other guys joined me at the window as we collectively resigned ourselves to fate. At least by the time they found out who did it, Kash would be gone.

You can't outrun the long arm of the Army, not as long as you're in it. But destiny is fickle. Remarkably, no valuable information was gleaned from the investigation—at least none that incriminated us.

As for my buddy Kash, like all the other friends I knew in the service, I never did see him again.

*Dear Handsome Harvey,*

*You should see how short I'm getting . . . I can barely see over my boots! My ETS—Elvis Taco Sandwich—order is up and on the counter. I can smell it.*

*I don't want you to think this is just the beer talking, but you should know, in spite of everything, you are and always will be my brother and I love you. (Forgive and forget is something else entirely.) I've made some really good friends, brothers from another, since I've been here, and I've seen them leave. I realize we knew each other through some of the worst shit and had some of the best times, but we never learned the details of each other's lives . . . like it was a boundary we didn't go beyond.*

*Take my buddy Kash . . . I have no idea who his parents are or where he came from. The most I know about Pug is that his family was from Philly, that his dad was a lifer and he is too. Big Ben's from Asbury Park, and that's all I know. Same with Roland from Idaho. I don't know any more about Crazy than his name. Same with Merrick, Washington, and the all guys from HQ.*

*You and me, we shared the same room, and sometimes the same bed, the same little tent and rowboat during the summers. The first time I got high was with you. Remember when we were brought home by the Mounties? And in Oak Park we got busted in the alley and I had to run from the cops because I was holding? And when we got thrown out of the theater by the Chicago cop for getting stoned during Up In Smoke? And when the cops "confiscated" our beer right outside the liquor store—and we could see the pizzas in their back seat. (I knew it was bullshit 'cause there was no room for us). I'm just saying . . . we have too much history to ever be strangers.*

*Claudine told me she was pregnant. She hasn't started to show, but I can't help but wonder if it's for real. I think she's just trying to keep me here with her, even though I've already told her I'm not ready for that kind of commitment. I haven't learned shit in the Army. I've got no education and nothing lined up. I'm not ready to be a father, and I'm definitely not staying in this country.*

*So it could be real, but this'll be her fifth kid from five different men. Fuck! I feel terrible for Claudine, but she knows I won't stay here . . . and she won't leave Germany. She gets a welfare check each month that lets her stay home with the kids. This is where her brot is buttered.*

*Honestly, how could I be a parent when I really don't know where I'm going or what I'm going to do? And I feel scared. I'm just so unprepared . . . I didn't gain any skills I can use on the outside. I need to go back to school, but I'll be 23 years old and I'm not even sure what I want to study. Remember when Fred Flintstone went back to college? That's how I see myself—except I don't think I'll fit in nearly as well.*

*I'm not going to have much money when I get out, no place of my own, no car, and no job to pay my way. Sure, I'll stay with the folks for a few weeks . . . till we drive each other crazy. And I'll crash on your couch for a week or so, until you kick me out. But then what? I'm going to miss enrollment for the spring semester and I'll have to do something till fall.*

*My GI Bill doesn't kick in until I'm registered, so that's a rubber biscuit that doesn't bounce back for half a year. So there's no way I can shoulder the responsibility of a family—and she's already got 4 kids! Leaving those kids is going to be harder for me than leaving her. And I may never know if she has my child or not.*

*How's that for a mind-fuck?!!*

*This should be my last transmission unless I get held up. The Big Green Weenie always has the last word. Hopefully I'll finish clearing all my paperwork by Tuesday and have a day off before I ship out Thursday. Regardless of my trepidation, I'm still thrilled to be getting the hell out of here. The thought of not being a chained dog brings tears to my eyes. Do you know what it's like living in such a controlled environment that even the color of your underwear is regimented? How I shave, how I cut my hair, even the look on my face has all been up to someone else.*

*Well, one thing in this world is for sure: the Bears are going to the Super Bowl and we are going to party like its 1999 in Chi-town. Yeah, Baby! Some places have dynasties and some places have franchises, but in Chicago we have the Junkyard Dogs. Bring it on!*

*On that optimistic note, I'll wrap it up. See you in a few weeks tops.*

*Love, Adam.*

# THIRTY

## Elephants Tiptoe Softly

In the military, there are limited ways of legally divorcing yourself from Uncle Sam without disgrace. Expiration Term of Service (ETS), most commonly known as Echo Tango Suitcase, is still as sweet regardless if it's called Exit To Safety, End The Sorrow, or Escape This Shit. Those magical initials are so much more desirable than their other far more tragic and better-known three-letter siblings: KIA, MIA, and POW. Uncle Sam calls ETS a "separation," because it takes at least six full years from your date of entry before he pulls the last of his meat-hooks out and you're truly single again. In the military, its not over until they say it's over—and not a minute before.

The Army is a people-mover bound with red tape; personnel are constantly rotated so no one gets to sink their roots too deep. And every time someone is transferred, there are always copious clearance forms to be filed, inventory to be returned, and waivers to be signed. The essential little black briefcase required to hold all my paperwork said "Short-Timer"—in other words, "leave this one alone."

I have always been an acute student of human behavior, and

when Sgt. Grimes started seriously threatening me with CCF with only one month to go, I headed to the PX for my golden ticket. They didn't need to know the cheap black case didn't hold any transfer paperwork, that it was merely full of books and magazines, and that my out-processing wouldn't start for another two weeks. All anyone needed to perceive was that I was on my way to being someone else's headache and not worth their time of day. I had watched enough short-timers to know what to do, so I went through the motions of pretending to be an untouchable.

Most of the equipment checks and recordkeeping were done at the big Pioneer Kaserne, a twenty-minute taxi ride away from the François Post. Every day, right after morning formation, I would head off-post. It didn't matter where I went as long as I was seen leaving by taxi. I would usually just go around the block to my girlfriend's place. Anytime I was questioned about where I was going, I would just shake my briefcase at the inquiring sergeant and state, "The Big Green Weenie's fucking me around again."

The great thing about Army bureaucracy is that you're not dealing with rocket scientists, just a lot of unmotivated people dealing with a lot of redundancy for very little pay. In such a system, apathy and ignorance run rampant. The resulting fog of ambivalence made it easy to get lost, for better or worse. I also had a lot of gear to replace before I checked out; whatever I couldn't lay my hands on I'd have to pay for out of pocket. It's not like the Army was going to say, "Thanks so much for sleeping outside in a freezing fart-sack for the last two years—why don't you keep it?"

So the sleeping bag I turned in was Kuntz's. I didn't consider it stealing because I took it right in front of him and then dared him to do something about it. I never forgave him for nearly killing me and everyone else up and down the gun line that day. I knew Kuntz had two bags anyway, and he'd rather spare one to keep all his teeth.

Among other things, I also needed to replace a big-ticket item—my gas mask—and that was provided via a careless second lieutenant who left his in the latrine. Live and learn, he'd be more careful with it in the future, and the lesson could save his life one day. At least the lieutenant made enough money to buy a new one. Privates aren't paid nearly as well.

I did want to take home a memento, something more along the lines of *Midnight Express* than a pair of German lederhosen. We had all heard the stories of the guys who got caught with contraband at the airport, but we had never heard about the guys who made it through. So I cooked up what I thought was a foolproof scheme: taking home five fifty-mark sticks of black Afghani hash. Strictly for medicinal purposes, of course—and for bragging rights if I made it through security.

With only two weeks to go, I was put on CQ duty every other day until I was to ship out. Maybe my top sergeant was trying to keep me out of trouble, but staying up for twenty-four hours every other day is hell on the system. Days in between were slept away with Claudine. We both knew our time was almost up and wanted to be together as much as possible. She knew I couldn't stay but hoped I'd come back. So every other day for the next fourteen days I manned the front desk around the clock and cleaned the first floor at night. Only after I woke everyone up the following morning could I go to sleep—but not until I had completed my paperwork.

During the nights when I swept, mopped, and buffed the tile floor to a reflective shine, I grew more resentful. I wanted something in return for being treated like shit.

In the hallway leading to the top sergeant's office, there was a large wooden display case with a simple lock guarding long-forgotten trophies. Pushed to the back was a dull pewter cup, about a foot tall, with an ornate cap. The engraving read: "Best Battery, Organization Day 1971, A Battery." Its simplicity appealed to me, and who the hell remembered 1971?

At three o'clock in the morning, before buffing the floor around it, I jimmied the lock of the case with the tweezers and toothpick from my Swiss Army knife. I removed all of the trophies, wiped the dust from the shelves to erase any "footprints," then put them all back—minus one. It was a gamble, but I was betting that no one paid enough attention to notice one missing. I couldn't have imagined that in less than six years, the first battalion, including the Death Dealers, would be disbanded. The fortieth field artillery regiment would be dissolved, and the entire third armored division, including that old trophy case and its contents, would cease to exist.

———

The day before shipping out, with all my out-processing taken care of, I met up with my dealer and picked up the hash. Then I went to downtown Hanau for the last time. I bought a gift set of five generously sized shot glasses from the Hinterland, along with a large red cinnamon-scented candle.

I carefully wrapped each piece of hash in plastic and set each one in a glass filled a quarter way with warm wax. Then I dripped enough wax over top to cover each stick with a thick layer. I finished each "candle" off with a short snippet of twine stuck into the top to resemble a wick, then repacked them in their box. I figured that the wax would x-ray solid and would hopefully block any aromas from the keen noses of the security dogs.

All my buddies were still on alert, so I had to leave without saying goodbye.

After checking in at the airport, the two hours I had to wait were nearly unbearable. I was sure I was going to get caught. Either they would spot the metal trophy in my baggage as it went through the x-ray machine, or the dogs would pick up the scent of gooey black Afghani. Perspiration pooled in dark circles under the arms of my jacket as I kept waiting to hear my name called over the loudspeaker to report to baggage claim, or for a squad of security guards to head my way. Knowing I looked guilty as hell, I spent the last forty-five minutes sweating it out in the bathroom, as if it was some sort of sanctuary. When I finally heard my flight announced, I expected to be grabbed from all sides as I stepped out. But lo and behold, I made it to Ft. Dicks, New Jersey, without a hitch.

Ft. Dicks was the GI equivalent of Ellis Island, where every soldier in the course of leaving the service passed through. It was the last time I would ever wear my BDUs and those black flat-footed boots, with just forty-eight hours to complete the last of my forms for final out-processing. Then, after two long years, I'd be homeward bound.

They gathered all of us two-year Minute Men into a large conference hall for debriefing. There were a lot of us and the atmosphere was full of nervous energy. We were given coffee, good coffee, with real cream and a cornucopia of doughnuts: big bear-claws like you can only find in the States, oversized Texas glazed, Vermont glazed maple, apple fritters. We had never been treated so well, and we were happy, which doesn't happen a lot in the Army. I couldn't help but think Uncle Sam, like always, had something up his sleeve.

Turns out I was right.

The colonel in charge, a favorite-uncle sort, gently lowered the boom on us. "Read the fine print, ladies and gentlemen. Six years is your minimum commitment to the US Army."

We weren't two-year men as we all thought. Tensions were running high in the Middle East and the brighter ones of us thought we knew what that meant.

"Four more years is what you owe the Red, White, and Blue," the colonel declared.

We glanced at each other in shock. No one had prepared us for this, but there was no protestation either. We all knew we had no basic rights. We were pieces of equipment at best and the Army could do with us what it wanted. It was built into our contracts.

The colonel had a solution, though. If we joined the National Guard or Army Reserve for just one weekend a month, we would continue to draw a salary and gain rank. Furthermore, not *if* but *when* each of us was recalled, only those in either the National Guard or Army Reserve would have their civilian jobs safely held for them until they got back home—guaranteed. Otherwise, we would simply get yanked back into service at Uncle Sam's precarious whim. When we finally got back out, there would be nothing waiting for us except the wind.

I could see people nodding in agreement. They were buying into it and I felt one last surge of patriotism while wearing the uniform. I put up my hand.

"Yes, son?" the kindly officer said.

"Sir, all I have to do is give up one weekend a month?" I asked incredulously.

"That's right, my boy."

"Well, sir," I boomed, standing solidly, "I only get four weekends in a month, and twenty-five percent is way too much to ask for. I'm going to take my chances and if they recall me, they're going to have to take me kicking and screaming, sir!"

I don't know who was more stunned, Uncle Colonel or his scared-straight audience.

It took the officer a moment to readdress me. "Son," he said, clearing his throat, "you can leave right now."

When a Full-Bird says that to a private, what he means is: Get the fuck out!

I had fully expected to be put on some shit detail—good old Army discipline for smart-alecks—for the next two days for opening my big mouth. But that colonel was a lot meaner than I thought possible.

It's a cruel joke to deny a person their future by taking away their past, but that's precisely what the military did. They fast-tracked my paperwork and out-processed me right then and there, two days shy of my full commitment. In the space on my DD214 where my signature should have been was typed "unavailable." In other words, the form upon which all my future benefits would be determined was incomplete, as was my two years of active duty.

In the military details matter. You don't think about those things when you're young, and that "clerical error" wasn't caught until twelve years later, when I was married and tried to buy a house using my GI benefits. In fact, the VA would use the lack of information in my files to deny me benefits related to any and all of my injuries, stating that none of them were "service related."

So the storm-burst headaches that sprung from my busted brow, the bouts of severe internal bleeding I suffered from being stomped, and my worsening limp from the crushed foot were all considered unsubstantiated. When the last three discs in my lower back spontaneously ruptured, just shy of my fortieth birthday, the VA told me there was no documentation and therefore no relevance. The fact that I was encouraged to run with four hundred pounds of TNT on my back was immaterial; none of the damage done to my body could possibly be their responsibility. As far as the VA was concerned, I was shit out of luck.

P.S. The Best Battery trophy cup is now property of the Field Artillery Museum located in Fort Sill, Oklahoma.

STATE OF ILLINOIS

OFFICE OF THE GOVERNOR

SPRINGFIELD 62706

JAMES R THOMPSON

GOVERNOR

February 6, 1986

Adam L. Harris

Dear Mr. Harris,

   Serving your country is an honor. As Governor, I
join with the 11 million people of Illinois to say,
"thank you for your service and welcome home."

   Much has been said and written about the honor of
serving in the military. But General Omar Bradley
perhaps best described the feeling of working with your
countrymen. "Our military forces are one team — in the
game to win regardless of who carries the ball," he
said. "Each player on this team — whether he shines in
the spotlight of the backfield, or eats dirt in the
line — must be an ALL American."

   It's that kind of spirit that places the life of the
military high in our society. In Illinois, we
appreciate the dedication that you have shown and the
commitment you have made in serving our country.

   The Illinois Department of Veterans' Affairs knows
what that commitment takes and stands ready to assist
you as you return to civilian life. Our representatives
at the local field office will be glad to serve you;
and I encourage you to call on them with your
questions.

   Again, thank you for your service and welcome home.

                    SINCERELY,

                    JAMES R. THOMPSON

                    GOVERNOR

# EPILOGUE

⌐⌐

I didn't fully realize how crazy the Army had made me until I got home. I had flown in from Frankfurt and got shot out of Ft. Dicks without a hot or a cot. By the time I got to my parents' house, the one they moved to while I was away, I was asleep on my feet. I awoke the next morning in a bizarro universe of comfortable clutter: throw pillows everywhere, walls lined with overflowing bookshelves, houseplants spilling out as if having sprouted overnight.

In the Army, I had gotten used to a very Spartan existence. My cot had defined my personal space; everything I owned fit neatly into a foot locker, and every facet of my life had taken on specific dimensions that didn't exist anywhere in the real world—and certainly not in my folks' home. Without a timetable and taskmaster, I didn't know what to do with myself. I was twenty-three years old and felt far behind the eight ball. I had no car, no job, no living arrangement, and no friends—because life and people move on without you. In return for my service and sacrifice, some sort of parachute to let me settle down nice and easy would have been really appreciated; it's not like being a private was such a lucrative position that after putting my life on the line for two years, I walked out with a stuffed wallet. Even guys getting out of the slammer have a halfway house, a job lined up, and someone to periodically check in on them. Not so when I left the Army. I just hit the ground with a plop.

For the first two weeks, I detoxed from all the military dogma by following a strict diet of beer, push-ups, and watching the news—thirty-six hours on, twelve off, round the clock. My folks stayed out of my way and let me be, which was a huge blessing. I knew my behavior was odd, but I didn't know how else to work it out of my system. I felt uncomfortable in their house, in the world around me, in my own skin.

When I started to venture out in my mum's little hatchback, I felt like I was looking at the world with my blinders removed. People were oblivious sheep selling their souls to the corporate dream store's shuck and jive. Get a job, buy a house, go into tremendous debt, and work to pay it off for the rest of your life. If you get sick, if you get fired, if you can no longer contribute—the system snatches everything away. America was the land of the people indebted to and enslaved by a system run by big money, from the lawmakers to the lawbreakers.

I felt like I was on the outside looking in, trying to find a spot to merge, with no roadmap or compass to guide me. I drove out to Ft. Sheridan thinking I could find people to relate to. I parked in the visitors lot and went into their PX, but my hair was growing out, my boots were still stuffed in my rucksack with my uniform, and I was no longer one of them. All my ties had been severed. I had no identity. I thought that maybe a veteran's group could help and my mother suggested keeping it in the tribe, so I looked to the Jewish War Veterans for support.

My biggest concern was that the Cold War wasn't considered a real war, regardless of casualties. That wasn't a problem for them—they rejected me because I wasn't an American and suggested that I contact a Canadian Jewish veterans group. I tried to explain that I had served for Uncle Sam, not the Canadian armed forces, and as such was a veteran of the US Army. But that didn't matter. The answer was still "not American enough." Maybe they were afraid I would dilute their ranks—that they would all start saying "Eh" and talk about hockey instead of baseball.

That's also when I realized that a Cold War veteran can't join the American Legion, nor the Veterans of Foreign Wars. Both are chartered by congress and are exclusive for combat veterans. Admission would require congress to rule that the Cold War was a war-time era. Being a Cold War veteran also meant being a third-class veteran, a sentiment echoed by the limited medical care the VA was willing to provide to non-combat veterans.

The ramifications of the lowered tier of medical care available to Cold War soldiers becomes a major concern to those veterans as they age. Some things, like teeth and eyes, are simply not covered because they are not considered "service related." So the only dentistry available for Cold War veterans are extractions, and at just over two hundred fifty dollars a pop, the VA will gladly pull a tooth. It didn't matter that I spent a lot of time in the field and was never issued a toothbrush or instructed on proper dental care while living outside. Teeth weren't integral to my job. Plain and simple, Cold War veterans receive the absolute minimum medical care because it saves money.

— — —

Orwell coined the term "Cold War" at the end of WWII, as if it was a single event. In reality, it was comprised of many military campaigns that raged throughout much of the world. These series of aggressions were promoted, funded, and fought between the United States and the Soviet Union using small countries and weak governments as the man-in-the-middle on a global scale. The resulting tug-of-war embroiled not only Southeast Asia, the Middle East, and South America but Africa and the Caribbean as well.

Whether these hostilities were called a war, a coup, or a police action depended on the expenditure involved and the media coverage they garnered. So, many combat situations where Americans were involved were ignored and downplayed: in Nicaragua it was called an "Intervention"; in Angola, a "Conflict"; in Laos, officially and stunningly truthfully, "The Secret War." The terminology is significant because only those who serve in a recognized war receive combat pay, along with all the benefits and honors that go with it.

The Cold War boiled and simmered from 1947 to 1991, but it was not a direct-combat engagement—or in other words, not a "real" war. Terrorists were considered "armed foreign forces," so the general wounded by the Red Army Faction in 1981, and the servicemen and women who were injured, some fatally, in bombings at the Frankfurt PX in 1985 and the West Berlin disco of 1986 all received Purple Hearts. The only other Purple Heart given during those Cold War years was to Maj. Arthur Nicholson, who in 1985 was shot and killed by a Soviet guard in East Berlin for spying. But none of the guys who got crushed or electrocuted or decapitated during constant maneuvers, not to mention those injured in a myriad of ways, ever had their sacrifices recognized—and none of their names were ever entered into the Book of Merit. Without official acknowledgment from the Defense Department, there would be no other medals given for all those who gave their lives and suffered grave injuries—nothing but a cold shoulder to match the war's moniker.

The difference in semantics goes beyond just what kind of ribbons get pinned to the green suit that hangs in the back of the closet. Cold War casualties and fatalities were not reported to the media, and documents were either never created or were destroyed. Nor are there statistics of how many Americans died or were wounded. History is written by the victors, and sometimes it's easier to just sweep people under the carpet and then burn the house down. So to this very day there are still veterans looking for witnesses to the accidents that occurred forty to fifty years ago because there's no documentation to stop the VA from turning down their claims. Not quite the "free healthcare" we were promised when we signed our lives away on that dotted line.

— — —

Once I realized the Army was no longer there for me, I had to reinvent myself and start over, and I felt that a change of scenery would do me some good. After a few weeks at my parents' house, it was time to move on. I found a company in the newspaper that needed a car driven from Chicago to Palm Springs, which was as close to Los Angeles as I could get, and said my goodbyes. Everything I owned was shoved into that little car, and I headed across the country not knowing what I would find when I got there. The company gave me six days to deliver the car, and I was able to make the drive in three.

I remember my first meal in California: pickled pigs feet, tabasco peppers, and a cold beer by the side of the road. It was eleven in the morning. I had pulled into a small service station after seeing orange trees growing along the freeway divider.

I knew what I was looking for even if I didn't know where I was going, and beginning at San Bernardino I sought out sports stores. I was searching for a surrogate platoon, and the day after landing in L.A., I found the Pomona Diablos Rugby Club. The next day I found a room to let in a dark little house that smelled of bad soup, owned by a weird, muu muu–wearing old lady. The following day, I drove the car to Palm Springs, dropped it off, and hitchhiked back to L.A. The very next day, I got my first job, pulling weeds.

After a miserable week of removing dandelions from endless banks of bee-buzzing ice-plants, I played my first rugby game for the Diablos and things changed. I had lost the necessary bulk to be an effective Prop and my right foot didn't allow me the ability to be the open field runner I used to be, but I more than made up for my shortcomings with off-the-chart aggression. The pitch promised to be a safe place for me to unload my anger issues. I was a hundred and sixty-five pound cannonball, knocking people over by the twos and threes. During the game, my new coach actually pulled me aside and told me to take it easy on the other team.

Afterward, at the bar, the stocky team captain came up to me. "We could use a good hooker around here," he said with a smile. "I hear you're looking for work?"

All I had to do was nod.

"Ben," shouted the captain, "can you give this guy a construction job?"

Next thing I knew, I was digging ditches and shoveling concrete, then moving my way up to labor foreman. "El Diablo Blanco" the workers called me. "Lead by Example" was my motto, and I was the diggingest dog they ever saw. We'd pile into the back of our boss's pickup to start each day before the crack of dawn, using each other for warmth. Without the sun, the desert's damn cold. By the end of the day, we'd ride back sunburned and exhausted from the heat. We'd get half a suitcase of cold Budweiser in the can if we had worked well—and we shoveled hard for that beer.

After a time, I moved into a ramshackle hut built for migrant workers during the Depression with three other ditch diggers. I was the only one who didn't speak Spanish and the only one who had a green card.

Early one evening, I called up the Alpha battery CQ desk after getting into the better part of a bottle of tequila. I was drunk enough to be homesick for the Army, and I missed my buddies. I figured Slim would already be transferred out, so I got the CQ to wake up Washington. It took a long while for Washington to get to the phone, and at first he acted like he didn't remember who I was. But after a few reminders, he blurted out, "Roland married Claudine . . . and they're gonna have a kid!"

"Wait? What? What about Roland's girlfriend?" I stammered. "Wasn't she pregnant? I thought they were going to get married."

"Yeah," he said. "They broke up. She married Roy the mechanic instead."

"Who the fuck is Roy?"

"He's Roland's best friend." I could tell Washington was getting pissed. "Uh, hey. It's three thirty and I'm going back to bed. Check the time difference before you call again."

The next thing I heard was a dial tone, buzzing over the one word that stuck in my mind: "kid."

That was *my* kid he was referring to. Mine.

I knew we would meet someday. I knew that as empirically true as I had known anything, ever.

———

After ten months of shoveling concrete, my brother Harvey came out to join me. He had been working in a TV station in Carbondale where he had interned while majoring in film. Harvey was going to be a director. I didn't know what I wanted to be, but I was hoping to put manual labor behind me. I knew I had to go back to school and get a degree, but I didn't know what to study. All I knew is that I wanted to live a Bohemian student's lifestyle—the young adult experience I had missed out on.

Harvey and I found an apartment right on Hollywood Boulevard, and together we moved west to the City of Broken Dreams. Harvey got a job working for Bozo the Clown, mostly copyright stuff, and I became a chimney sweep in sunny Southern California—with a top hat and a tuxedo T-shirt. How crazy is that?

I never again saw any of the guys I served with, and I have always wondered what happened to them. They were the best people, and the best friends, I ever had. I especially wonder about Pug, if he ever tells stories about us. In my fantasy, I see all of them as successful family men who are loved, respected and well off . . . all the things I would have wished for myself but never quite found my way to achieving.

"Whatever God does is for the best."

–Rabbi Akiva, 2nd century CE

Lookin' Like Bambi Caught in the Headlights
(I think my lips are so pushed out because the chin strap is too tight!)

Basic Training 3rd Platoon (guess who's holding the flag?)

Awarded Most Distinguished Graduate

Last Day of Basic at Ft. Sill

My Home Sweet Home

A Howitzer and Its Gun Bunny

Not the Most Savory Invitation for a Meal . . .
"Who wants to eat at the Death Dealers tonight?"

Flexing the Guns

The Golden Nugget Saloon AKA My Office

Find the Rugby Player on the Soccer Team

The Javelin Throw That Almost Got the Ump

Advance Party Chilling in an Onion Field
Just Klicks from the Czech Border

Me and Mr. Robinson

Before Going Out on the Town

Pug and I: The Immortals Go Skiing

Remember the 80s?

Closest I Ever Came to Born in the USA

The Canadian?
He's the One Smiling in the Snow

Top Hat, Tuxedo T-Shirt and Rugby Shorts:
Standard Hollywood Chimney Sweep Issue

The Fuse Setter

## BOOK DISCUSSION QUESTIONS

1.  If you have served in the military, how did you relate to Adam's experience? How was yours different?

2.  What was your reaction to Adam's being left behind by his unit after his horrific foot injury? How do you think it should have been handled?

3.  Did you think Adam's sometimes hostile reactions to his military superiors were justified? Do you believe that we should stand up for ourselves, no matter the circumstances? If no, why not?

4.  How did Adam's letters home to his brother enhance the memoir for you? Did they offer an alternative perspective on his experience at times?

5.  Adam endured several brutal assignments, pushing human tolerance to the limit and even to the point of cruelty. Do you believe this is justifiable treatment for service members? Why or why not?

6.  What is your opinion of the ways that Adam, as a young man in his early twenties, coped during his conscription (alcohol, women, occasional drug use)? Were any surprising? If so, why?

7.  Adam suffered from several events of brutality, some unprovoked, that left him with lasting injuries. What did you make of these violent acts against him?

8.  Adam played a lot of pranks and also got himself into some humorous situations. Which events evoked laughter? Shock? Cringing? Cheers?

9.  How do you think Adam's childhood as a marginalized kid influenced his behavior in the Army?

10. When Adam was just shy of release, he publicly refused to give any more time to the military, which resulted in jeopardizing his future benefits. What did you make of Adam's decision? Of the colonel's? Do you think that Adam standing on his convictions was admirable, despite the outcome for him?

11. What did you take away from Adam's story? Did your understanding of military sacrifice change? Of military treatment of soldiers, in general? What other elements were thought-provoking for you?

## ACKNOWLEDGMENTS

Without my first, and most enduring, hero, my grandfather, and without the one and only person to show me what unconditional love was, my grandmother, I don't know where I would be or who I would be today. Everything good that I have and all the very best parts of me, I owe to them.

I must also acknowledge and thank all the storytellers and writers who have influenced me, from Pete Seeger to Rosemary Sutcliff to Gerald Durrell and so many more to the present. You can blame *The Elegance of the Hedgehog* for giving me the courage to write this, but I have only thanks and admiration for Muriel Barbery.

Reading not only helped me escape throughout my childhood, but during the 80s, while in the service, books were all the entertainment we had as soldiers in the field. And to be fair, I must also thank the US Army for helping me mature and find my voice. In a way, you could say that field artillery helped me come out of my shell.

Lastly I'd like to acknowledge that because of, or perhaps in spite of, everything I've been through, I still don't know why I'm here kicking around. But maybe it was for this, for these stories to have a life of their own and perhaps to live forever through them.

# ABOUT THE AUTHOR

Adam Harris resides in the north-east of Ontario, Canada, way above the line of latitude where most in North America live, in a town so small that during the sixty seconds it takes to pass through, one would be forgiven for not noticing the one and only traffic light (which stays green all the time). Too remote for a mailman, the post office where he picks up his mail also sells hunting licenses  and shares its roof with the bait shop. His closest neighbor is a chipmunk whose precociousness is matched only by his cuteness and gluttony for nuts. Far less welcome, and fortunately far more timid, are the bears that seem to spontaneously spring from the surrounding forests.

With degrees in dietetics, nutrition, food science, and the culinary arts, as well as a background in restaurant work, Adam sees the world through food-colored glasses. The first question he asks himself every day is, "What's for breakfast?" followed shortly by "What's for dinner?" He started cooking at an early age and feels grateful to be able to prep and serve pretty much anything he wants from his favorite restaurant: his own kitchen.

Parts of Adam's resume read like something out of a Dickens' novel—mapmaker, cannon crewman, chimney sweep—while others range from ditch digger to Le Cordon Bleu assistant chef to senior consumer scientist. He's spent most of his life working abroad and is now happy to be home spending his days building furniture and tinkering with his model ships. And maybe, just maybe, if plied with enough whisky, he'll tell some more stories.

www.ingramcontent.com/pod-product-compliance
Lightning Source LLC
Chambersburg PA
CBHW030401130626
46549CB00004B/1589